Hassan Rezaei

Biosorption of Heavy Metals by Using Algae

Hassan Rezaei

Biosorption of Heavy Metals by Using Algae

Physical Chemistry Parameters of Biosorption

LAP LAMBERT Academic Publishing

Impressum/Imprint (nur für Deutschland/only for Germany)
Bibliografische Information der Deutschen Nationalbibliothek: Die Deutsche Nationalbibliothek verzeichnet diese Publikation in der Deutschen Nationalbibliografie; detaillierte bibliografische Daten sind im Internet über http://dnb.d-nb.de abrufbar.

Coverbild: www.ingimage.com

Verlag: LAP LAMBERT Academic Publishing GmbH & Co. KG
Heinrich-Böcking-Str. 6-8, 66121 Saarbrücken, Deutschland
Telefon +49 681 3720-310, Telefax +49 681 3720-3109
Email: info@lap-publishing.com

Herstellung in Deutschland:
Schaltungsdienst Lange o.H.G., Berlin
Books on Demand GmbH, Norderstedt
Reha GmbH, Saarbrücken
Amazon Distribution GmbH, Leipzig
ISBN: 978-3-8465-8193-3

Imprint (only for USA, GB)
Bibliographic information published by the Deutsche Nationalbibliothek: The Deutsche Nationalbibliothek lists this publication in the Deutsche Nationalbibliografie; detailed bibliographic data are available in the Internet at http://dnb.d-nb.de.

Cover image: www.ingimage.com

Publisher: LAP LAMBERT Academic Publishing GmbH & Co. KG
Heinrich-Böcking-Str. 6-8, 66121 Saarbrücken, Germany
Phone +49 681 3720-310, Fax +49 681 3720-3109
Email: info@lap-publishing.com

Printed in the U.S.A.
Printed in the U.K. by (see last page)
ISBN: 978-3-8465-8193-3

Biosorption of Heavy Metals by Using Algae

Dr. Hassan Rezaei

TABLE OF CONTENTS

NOTATION

b	Amount of adsorbed per unit weight of adsorbent (Langmuir constant)
BOD	Biochemical Oxygen Demand
COD	Chemical Oxygen Demand
C_0	Initial ion concentration
C_e	Equilibrium or Final ion concentration
C_a	Milligram of adsorbent adsorbed per liter
C_e	Equilibrium concentration of solution
°C	Degree centigrade
EC	European Community
EPA	Environmental Protection Agency
EDS	Electron Dispersive Spectroscopy
FTIR	Fourier Transform Infrared
IR	Infrared
ISX	Insoluble starch xanthate
K	Kelvin
K_c	Thermodynamic equilibrium constant
K_f	Equilibrium constant indicative of adsorption capacity(Freundlich constant)
K_2	pseudo-second-order sorption
L	Litter
M	Biomass dry weight
M	Molar
Min.	Minute
mg	Milligram
meq	Milli equivalent
mL	Milliliter
MBBM	Modified Bold' Basal Medium
n	Adsorption equilibrium constant(Freundlich constant)
NCIM	National Collection of Industrial Microorganisms
NMR	Nuclear Magnetic Resonance
pH	- Log $[H^+]$

ppm	Part Per Million
q	Biosorption ions uptake capacity
q_e	Biomass biosorption equilibrium ions uptake capacity
q_{max}	Maximum ions uptake capacity
q_t	Amount of metal ion adsorbed at any time
R	Ideal gas constant
R^2	Coefficient of determination
R_L	Equilibrium parameter
RS	Raman spectroscopy
rpm	Revolutions Per Minute
SEM	Scanning Electron Microscopy
Sp.	Species
t	Time
T	Temperature in K
V	sample volume
XAFS	X-ray Absorption Fine structure Spectroscopy
XPS	X-ray Photoelectron Spectroscopy
XRD	X-ray Diffraction
$\Delta G°$	Gibbs free energy change
$\Delta H°$	Enthalpy change
$\Delta S°$	Entropy change

ABSTRACT

The present study attempts to understand to what extent biomass application can be useful to treat industrial effluent so as to isolate heavy metals. The researcher has carried out various laboratory experiments to identify suitable methods. The present report is documentable of the work discussion on previous literature, methodology and the relevance.

Heavy metals are present in nature and industrial wastewater. Due to their mobility in natural water ecosystems and toxicity, the presence of heavy metals in surface water and ground water has become a major inorganic contamination problem. Discharge and treatment of industrial wastewater containing heavy metals are important issues in environmental protection. Various industries produce and discharge wastes containing different heavy metals into the environment. *Biosorption is a property of certain types of microbial biomass to concentrate heavy metals from aqueous solution.* **This is new research to use of blue green and green algae, which are growing on automobile industrial effluent naturally and we can to use them for removal of the heavy metals.**

In this project we collected them and after identify of them, we understood three types of algae were on the effluent treatment plant that includes the *Oscillatoria sp.*, *Chlorella pyrenoidosa* **and *Spirulina sp.*. In this study after identify the algaes for laboratory study we collected pure algaes.** *Chlorella pyrenoidosa* (2738) and ***Spirulina sp.*** (5143) obtained from National Collection of Industrial Microorganisms (NCIM) from Pune - INDIA and *Oscillatoria sp.* was obtaining from agriculture collage from Pune - INDIA, which was isolated and thoroughly pure. The *Oscillatoria sp.*, ***Spirulina sp.*** and *Chlorella pyrenoidosa* were maintained in modified Bold' basal medium (MBBM), ***Spirulina sp.*** Media and Fog's Media at 28 ^{0}C with using 3000 lx light intensity respectively. The biomasses dried at 60 °C in an oven for 24 h and milled to a gritty consistency. The biomasses were sieve for particle size smaller than 1 mm and stored in dark bottle and keeps in a dry cabinet for experiments. All of the media are sterilized by autoclaving at 121°C for 20 min. In the pre-treatment, the biomass was slowly stirred in the $CaCl_2$ for a certain period of time.

In the batch study conical flasks (250 mL) were shaken at 120 rpm in a temperature controlled rotatory shaker and in the column study *3 g of dry biomass were packed into a 50 cm long column of 2.5 cm inner diameter made of stainless still.* Ions were determined by atomic adsorption spectrophotometer.

All of three biomasses (*Chlorella pyrenoidosa, Oscillatoria sp.* and *Spirulina sp.*) has good potential for remove a heavy metals from waste water and they can be used in industrial for remove the heavy metals. With increase the initial concentration percentage of biosorption decreased and metal ion uptake capacity was increased. With increase the pH up to 5 percentage of biosorption increased and metal ion uptake capacity was decreased and after pH=5 percentage of biosorption decreased and metal ion uptake capacity was increased. With increased the amount of biomass observed that percentage of biosorption increased and metal ion uptake capacity was decreased. Optimum temperature and pH was 25 °C and 5 respectively. All of the result showed that the nitric acid was the best chemical solution for desorption process.

The equilibrium experimental results of heavy metals have been fitted in the Langmuir and Freundlich models but coefficient of determination (R^2) Freundlich is bigger than of Langmuir and that means, Freundlich model is better model for our study. Equilibrium time was about 60 minute. The R_L values at different initial adsorbate concentrations indicate favorable adsorption for all the adsorbents and adsorbates studied. Parameter of kinetic model for heavy metals of biomasses showed that with increase the initial concentration coefficient of determination (R^2) and K_2 decreased.

A positive value of $\Delta H°$ indicated the endothermic nature of the process. A negative value of the free energy ΔG° indicated the spontaneous nature of the adsorption process. It was also noted that the change in free energy, increases with increase in temperature, which exhibits an increase in adsorption with rise in temperature. This could be possibly because of activation of more sites on the surface of biomasses with increase in temperature. A positive value of $\Delta S°$ showed increased randomness at solid–liquid interface during the adsorption of heavy metals, it also suggests some structural changes in the adsorbate and the adsorbent.

FTIR Spectrums of *Chlorella pyrenoidosa, Oscillatoria sp.* and *Spirulina* showed that revealed the presence of hydroxyl, amino, carboxylic and carbonyl groups. The presence of OH group along with carbonyl group confirmed the presence of carboxylic acid groups in the biosorbent. The presence of NH group and OH group along with carbonyl group might be attributed the presence of amino acid groups in the biosorbent.

Scanning electron microscopy (SEM) of *Chlorella pyrenoidosa, Oscillatoria sp.* and *Spirulina sp.* at before and after biosorption of chromium, copper, nickel and zinc clearly showed biosorption of heavy metals by algae biomass.

Chapter I: Introduction

1.1 Introduction

Heavy metals are present in nature and industrial wastewater. Due to their mobility in natural water ecosystems and toxicity, the presence of heavy metals in surface water and ground water has become a major inorganic contamination problem. Discharge and treatment of industrial wastewater containing heavy metals are important issues in environmental protection. Wang and Chen (2006) reported three kinds of heavy metals are of concern, including toxic metals (such as Hg, Cr, Pb, Zn, Cu, Ni, Cd, As, Co, Sn, etc.), precious metals (such as Pd, Pt, Ag, Au, Ru etc.) and radio nuclides (such as U, Th, Ra, Am, etc.). There are various methods to check the heavy metals pollution. One of them is biosorption. The present study attempts to understand to what extent biomass application can be useful to treat industrial effluent so as to isolate heavy metals. The researcher has carried out various laboratory experiments to identify suitable methods. The present report is methodology of the work along with discussion on previous literature, methodology and the relevance. This chapter is devoted to give brief introduction to the topic and related concept

1.2 Sources of heavy metal containing wastewater

With the development of industrialization and human activities, the discharge of waste and wastewater containing heavy metals to environment has increased. Various industries produce and discharge wastes containing different heavy metals into the environment, such as mining, surface finishing industry, energy and fuel production, fertilizer and pesticide industry and application, metallurgy, iron and steel, electroplating, electrolysis, electro-osmosis, leatherworking, photography, electric appliance manufacturing, metal surface treating, aerospace and atomic energy installation etc. Thus, metal as a kind of resource is becoming shortage and also brings about serious environmental pollution, threatening human health and ecosystem. In addition, domestic effluents, landfill leachate, agricultural runoff, and acid rain also contribute to heavy metals in wastewater (Aksu and Kutsal, 1990; Pradhan and Levine, 1992). Cr (VI), Cu, Ni and Zn are the heavy metals commonly found in such wastewater discharges. Permissible limited of heavy metals for discharge from the automobile industry as per US Environmental Protection Agency (USEPA) is on the Table 1.1 (EPA, 1980).

Table 1.1 Permissible limited concentration of heavy metals for discharge from automobile industry

Heavy metals	Cr (VI)	Cu	Ni	Zn
Permissible Concentration(mg/l)	0.1	3	3	5

1.3 Health impacts of heavy metals and regulation in control of heavy metals

When heavy metals are present even in a very low concentration, due to their recalcitrance and consequent persistence in water bodies, their concentration may be elevated through biomagnifications to a level that they start exhibiting toxic characteristics (Atkinson et al., 1996 & 1998).Thus, many countries set up guidelines and regulations for the discharge of wastewater containing heavy metals in order to prevent the impacts of heavy metal contamination on environment and humans. The standard metal concentrations in drinking water and their health effects as understood through literature summarized in Table 1.2.

1.4 Removal of heavy metals from wastewater

Treatment of industrial effluents containing heavy metals is mainly based on precipitation, ion exchange and activated carbon adsorption. Precipitation is based on the low solubility of heavy metal hydroxides or sulfides. Metal hydroxides and sulfides are formed by adding caustic soda or lime and hydrogen sulfide gas (H_2S) respectively to wastewater that contains heavy metals. The resulting heavy metal hydroxides or sulfides have very low solubility (Ku, 1987). The precipitated solid hydroxides or sulfides may then be separated by various solid-liquid separation techniques such as sedimentation or filtration. However, there are some limitations to the precipitation method. When the concentration of the contaminant is low, the precipitation reaction is correspondingly very slow. As a result, an increase in the consumption of lime or caustic soda occurs and very large mixing and decanting tanks would be required. Therefore the removal of heavy metals by precipitation from diluted effluents is not economical.

Table.1.2 The standard metal concentration in drinking water and the health

METALS	EFFECT	DRINKING WATER STANDARDS

Chromium	Necrosis • nephrites and death in man (10 mg/ kg of body weight as hexavealent chromium) Irritation • of gastrointestinal mucosa	By • the Environmental Protection Agency Maximum con.: (hexavealent and Trivalent) Total 0.1 mg/L By • European Community (EC): 0.5 mg/L Regulation • of water quality 0.1 mg/L
Copper	Causes • damage in a variety of aquatic fauna Phytotoxic • Mucosal • irritiation and corrosion Central • nervois system irritation followed by depression	By • the Environmental Protection Agency Maximum con.: 1.0 mg/L By • European Community (EC):3 mg/L Regulation • of water quality 0.01 mg/L
Nickel	High • conc. can cause DNA damage Eczema • of hands High • phytotoxicity Damaging • fauna	•By the Environmental Protection Agency Maximum con.: 0.1 mg/L By • European Community (EC): 0.1 mg/L Regulation • of water quality 0.1 mg/L
Zinc	Phytotoxic • Anaemia • Lack • of muscular coordination Abdominal • pain etc.	By • the Environmental Protection Agency Maximum con.: 5 mg/L By • European Community (EC): 5 mg/L Regulation • of water quality 0.1 mg/L

Furthermore, precipitation processes will generate toxic sludge, which has to be disposed of the final disposal of sludge is becoming more strictly regulated and corresponding disposal costs are increasing very rapidly. Purification of water supplies by activated carbon adsorption has a short history compared to other purification processes.

The first granular activated carbon units for treatment of water supplies were constructed in Hamm, Germany in 1929 and in 1930 at Bay City, Michigan (Montgomery, 1985). Its use became widespread in the next few decades, for the removal of organic components from aqueous solution. Nowadays, adsorption on activated carbon is a recognized method for the removal of heavy metals from wastewater. According to Ku (1987), activated carbon was used as a polishing step following either metal hydroxide or metal sulfide precipitation in the removal of residual zinc and cadmium from plating wastewaters. Similar to carbon sorption, the ion exchange process is usually implemented in a continuous-flow column packed with cationic and/or anionic resins. The packed bed reactor is effective for processing high volumes of effluents with low metal concentrations. The metals would sorb onto the resins in exchange for other metal cations or protons originally present on the ion exchange sites releasing them into the solution. Since heavy metals uptake at low pH is very low, the bound heavy metal cations may be simply eluted from the resin using a low-pH wash (Greene, et al., 1987; Guibal, et al., 1992; Treen-Sears, et al., 1984; Tsezos, 1980). This would make it possible to recover heavy metals in the form of liquid concentrates. An advantage of using ion exchange is that heavy metals may be sequestered from wastewater selectively when the appropriate resin is chosen. Many ion exchange resins are commercially available for removal of various metals (Anonymous, 1989). Such selectivity is especially useful to recover valuable heavy metals such as gold and silver. However, the higher the selectivity, the more strongly the metals are bound to the resin, and the more difficult it is to desorb them. This would increase the operating cost as more regenerants are used. There are other limitations to the application of ion exchange resins. Chemical degradation can occur with breaking the polymer network or modification of functional groups by a species in the solution. Also organics often cause fouling of the resin particularly gel-type resins (Harland, 1994). The most serious limitation of ion exchange is that ion exchange resins are often expensive in large-scale projects. Therefore, in order to keep the operating cost of the ion exchange process low, it is applied only to effluents with medium or low levels of heavy metals Heavy metals can be also removed by using biological treatment systems. Activated sludge technology is used in most industrial countries to process municipal sewage. However, the resulting bio-sludge is highly toxic and needs to be disposed of. The toxicity of this sludge prevents its use as agricultural fertilizers. Disposal by incineration causes another form of pollution in addition to further increasing sludge disposal cost.

1.5 Biosorption technology

Biosorption is a property of certain types of microbial biomass to concentrate heavy metals from aqueous solution. Only within the past decade has the potential of metal biosorption by biomass materials been well established. It was the work of Lowitz in 1785 that pioneered the field of ion exchange which eventually lead us to the field of biosorption (Montgomery, 1985). Biosorbents could be industrial waste biomass, as is the case in using byproduct biomass from fermentation processes, or naturally occurring substances of biological origin such as algae, which is found in large quantities in the sea (Volesky and Holan, 1995). It is particularly the cell wall structure that is principally responsible for this phenomenon. In recent years biosorption has been proposed to be a safe and cost effective process for the removal of toxic heavy metals from dilute metal-bearing solutions. Compared with conventional methods for removing toxic metals from industrial effluents, the biosorption process offers several advantages such as the efficiency of biosorption removal of toxic heavy metals is high. Effluents qualities in the order of only ppb (mg/L) of residual metals can be achieved (Volesky, 1999). Second, biosorption may be implemented over a wide range of operating pH, pressures and temperatures.

1.6. Advantage and Characteristics of biosorption

Biosorption has some advantages compared with conventional techniques (Volesky, 1999) as listed below:

- **Cheap:** The cost of the biosorbent is low since they often are made from abundant or waste material.

- **Metal selective:** The metalsorbing performance of different types of biomass can be more or less selective on different metals. This depends on various factors such as type of biomass, mixture in the solution, type of biomass preparation and physicochemical treatment.

- **Regenerative:** Biosorbents can be reused, after the metal is recycled.

- **No sludge generation**: No secondary problems with sludge occur with biosorption, as is the case with many other techniques, for example, precipitation.

- **Metal recovery possible**: In case of metals, it can be recovered after being sorbed from the solution.

- **Competitive performance**: Biosorption is capable of a performance comparable to the most similar technique, ion exchange treatment. Ion exchange is, as mentioned above, rather costly, making the low cost of biosorption a major factor.

Biosorbents intended for bioremediation environmental applications are waste biomass of crops, algae, fungi, bacteria, etc., which are the naturally abundant.

Biosorption by microorganisms have various disadvantages, and hence many low cost adsorbents (industrial/agricultural waste products/byproducts) are increasingly used as biosorbents.

1.7 Usage of biomass algae for biosorption of heavy metals from automobile industrial effluent

This may be the first attempt its kind to use of blue green and green algae, which are growing on automobile industrial effluent naturally and we can collect them and cultivate and after drying that, we can to use them for removal of the heavy metals. In this project we collected them and after identify of them, we understood three types of algae were on the effluent treatment plant that includes the *Oscillatoria sp., Chlorella Pyrenoidosa* and *Spirulina sp.* scientific classification of three algae summarized on the Table.1.3.

Table.1.3 Scientific classification of algaes

Scientific classification	Type of Algae		
	Chlorella Pyrenoidosa	*Oscillatoria sp.*	*Spirulina sp.*
Kingdom	*Plantae*	*Bacteria*	*Bacteria*
Division	*Chlorophyta*	*Cyanobacteria*	*Cyanobacteria*
Class	*Trebouxiophyceae*	*Cyanophycease*	*Cyanophycease*
Order	*Chlorellales*	*Oscillatoriales*	*Oscillatoriales*
Family	*Chlorellacease*	*Oscillatoriacease*	*Oscillatoriacease*
Genus	*Chlorella*	*Oscillatoria*	*Spirulina*

1.8 Objectives and scope of the study

The main objectives of this project were physical chemistry study of biosorption that were as follows:

1. Study of effect of pH, contact time, temperature, biomass concentration.
2. Study of batch and continuous sorbent treatment process.
3. Study of chemical properties of sorbent.
4. Study of adsorption behavior, adsorption kinetics and adsorption isotherm.

Chapter II: Literature of Review

2.1. Introduction

The toxicity due to heavy metals has been a concern for environmentalists and planners. The issue has become significant as the level of development. Several scholars like Esposito et al. (2001) have clarified that heavy metals are toxic pollutants released into the environment as a result of different activities such as industrial, mining and agriculture. Using the adsorption process for the removal of heavy metals from wastewater has a short history compared to other water purification processes (Hammaini et al., 1999). Biosorption is the ability of certain types of microbial biomass to accumulate heavy metals from aqueous solutions. Microbial biomass could be considered as an ion exchanger of a biological origin (Volesky, 1999). A large number of micro-organisms belonging to various groups, such as bacteria, fungi, yeasts and algae have been reported to bind a variety of heavy metals to some extents (Volesky, 1990). The present work focuses on heavy metals contamination in the effluent of automobile industry. This chapter mainly attempts to present a brief survey of literature related to how the issue can be resolved by adopting biosorption methods.

2.2 Heavy metals

Chromium water pollution has become of considerable concern due to the fact that chromium has been widely used in metal finishing, electroplating, leather tanning, stainless steel production, textile industries, and chromate preparation (Abdel-Jawad *et al.* 2002). Chromium exists in the environment in either its hexavalent form or trivalent form. The metal species Cr (VI) is considered as highly toxic in that it could act as a carcinogen, mutagen, and teratogen in the biological system (Abu Al-Rub *et al.* 2002).It has also been noted that prolonged exposure to the metal species could cause skin allergies and cancer in human beings (Ahalya and Ramachandra , 2002). Additionally, chromium can be oxidized to the state of a more carcinogenic and mutagenic Cr (VI) by some bacteria in the environment under certain conditions (Ahalya *et al.* 2006). Recently, the commonly used methods applied to remove excessive copper from aqueous solutions have included ion exchange, chemical precipitation, activated carbon adsorption, evaporation and membrane processes. However, these methods were found to be either inefficient or expensive when metal ions exist in low concentrations (<100 mg/L) and may also be associated with the generation of secondary environmental problems from waste disposal (Ahalya *et al.* 2003).

Ni (II) is present in the effluents of silver refineries, electroplating, zinc base casting and storage battery industries (Sitting, 1976, Al-Asheh and Duvnjak 1997; Kadirvelu, 1998; Beliles1979).Zinc is an essential element as enzyme activators in humans, but it is equally toxic at levels of 100–500 mg/day and it is a known carcinogen (Volesky and Holan, 1995).However, these methods were found to be either inefficient or expensive when metal ions exist in low concentrations (<100 mg/L) and may also be associated with the generation of secondary environmental problems from waste disposal (Ahalya *et al.*, 2003).

2.3 Removal of heavy metals

Continuous discharge of industrial, domestic and agricultural wastes in rivers and lakes causes deposit of pollutants in sediments. Such pollutants include heavy metals, which endanger public health after being incorporated in food chain. Heavy metals cannot be destroyed through biological degradation, as is the case with most organic pollutants. Incidence of heavy metal accumulation in fish, oysters, mussels, sediments and other components of aquatic ecosystems have been reported from all over the world (Naimo, 1995; Sayler *et al.*, 1975).

2.3.1 Chemical precipitation

Precipitation of metals is achieved by the addition of coagulants such as alum, lime, iron salts and other organic polymers. The large amount of sludge containing toxic compounds produced during the process is the main disadvantage. Following lines are devoted to outline different process to remove heavy metals as stated by different scholars.

- **Hydroxide precipitation**

Chemical precipitation of heavy metals as their hydroxides using lime or sodium hydroxide is widely used. Lime is generally favored for precipitation purposes due to the low cost of precipitant, ease of pH control in the range of 8.0 –10.0 and the excess of lime also serves as an adsorbent for the removal of metal ions. The limitations of this method include difference between metals in the optimum pH for hydroxide formation may lead to the problems in the treatment of effluents containing combined metal ions. Variability in metal hydroxide solubility at a fixed pH is another drawback.

- **Carbonate precipitation**

Carbonate precipitation of metals using calcium or sodium carbonate is very limited. Patterson *et al.*, 1997 reported improved results using carbonate precipitate for Cd (II) and Pb (II) from electroplating effluents.

- **Sulphide precipitation**

Since most of the heavy metals form stable sulphides, excellent metal removal can be obtained by sulphide precipitation. Treatment with sulphides is most advantageous when used as a polishing step after conventional hydroxide precipitation or when very high metal removals are required.

2.3.2 Chemical reduction

Reduction of hexavalent chromium can also be accomplished with electro-chemical units. The electrochemical chromium reduction process uses consumable iron electrodes and an electric current to generate ferrous ions that react with hexavalent chromium to give trivalent chromium as follows (USEPA, 1979).

$$3Fe^{2+} + CrO4^{2-} + 4H2O \rightarrow 3Fe^{3+} + Cr^{3+} + 8OH^{-}$$

Another application of reduction process is the use of sodium borohydride, which has been considered effective for the removal of mercury, cadmium, lead, silver and gold.

2.3.3 Xanthate process

Insoluble starch xanthate (ISX) is made from commercial cross linked starch by reacting it with sodium hydroxide and carbon disulphide. To give the product stability and to improve the sludge settling rate, magnesium sulphate is also added. ISX works like an ion exchanger, removing the heavy metals from the wastewater and replacing them with sodium and magnesium. Average capacity is 1.1-1.5 meq of metal ion per gram of ISX (Anon, 1978).

2.3.4 Solvent extraction

Liquid-liquid extraction (also frequently referred as solvent extraction) of metals from solutions on a large scale has experienced a phenomenal growth in recent years due to the introduction of selective complexing agents.

2.3.5 Membrane process

Important examples of membrane process applicable to inorganic wastewater treatment include reverse osmosis and eletrodialysis (EPA, 1980). These processes involve ionic concentration by the use of selective membrane with a specific driving force. The application of the membrane process described is limited due to pretreatment requirements, primarily, for the removal of suspended solids. The methods are expensive and sophisticated, requiring a higher level of technical expertise to operate. A liquid membrane is a thin film that selectively permits the passage of a specific constituent from a mixture.

2.3.6 Evaporators

In the electroplating industry, evaporators are used chiefly to concentrate and recover valuable plating chemicals. Four types of evaporators are used throughout the electroplating industry (USEPA, 1979a) (I) Rising film evaporators; (II) Flash evaporators using waste heat; (III) submerged tube evaporators; (IV) Atmospheric evaporators. Both capital and operational costs for evaporative recovery systems are high. Chemical and water reuse values must offset these costs for evaporative recovery to become economically feasible.

2.3.7 Cementation

Cementation is the displacement of a metal from solution by a metal higher in the electromotive series. Due to its low cost and ready availability, scrap iron is the metal used often. Cementation is especially suitable for small wastewater flow because a long contact time is required. Some common examples of cementation in wastewater treatment include the precipitation of copper from printed etching solutions and the reduction of Cr (VI) in chromium plating and chromate-inhibited cooling water discharges (Case, 1974).

2.3.8 Ion exchange

Ion exchange resins are available selectively for certain metal ions. The cations are exchanged for H^+ or Na^+. The cation exchange resins are mostly synthetic polymers ontaining an active ion group such as SO_3H. The natural materials such as zeolites can be used as ion exchange media (Van der Heen, 1977). The modified zeolites like zeocarb and chalcarb have greater affinity for metals like Ni and Pb (Groffman *et al.*, 1992). The limitations on the use of ion exchange for inorganic effluent treatment are primarily high cost and the requirements for appropriate pretreatment systems.

2.3.9 Electro deposition

Some metals found in waste solution can be recovered by electro deposition using insoluble anodes. For example, spent solutions resulting from sulphuric acid cleaning of Cu may be saturated with copper sulphate in the presence of residual acid. These are ideal for electro-winning where the high quality cathode copper can be electrolytically deposited while free sulphuric acid is regenerated.

2.3.10 Reverse Osmosis

It is a process in which heavy metals are separated by a semi-permeable membrane at a pressure greater than osmotic pressure caused by the dissolved solids in wastewater. The disadvantage of this method is that it is expensive.

2.3.11Bioremediation

Bioremediation can be defined as any process that uses microorganisms, fungi, green plants or their enzymes to return the natural environment altered by contaminants to its original condition. Heavy metal bioremediation involves removal of heavy metals from waste water and soil through metabolically mediated or physico-chemical pathways.

Algae, bacteria and fungi and yeasts have proved to be potential in metal removal from the waste waters (Volesky, 1986).

2.3.12 Adsorption

Since activated carbon also possesses an affinity for heavy metals, considerable attention has been focused on the use of carbon for the adsorption of hexavalent chromium, complexed cyanides and metals present in various other forms from wastewaters. Watonabe and Ogawa (1929) first presented the use of activated carbon for the adsorption of heavy metals.

2.3.13 Biosorption

Biosorption can be defined as the removal of metal or metalloid species, compounds and particulates from solution by biological material (Gadd, 1993).Biosorption is the binding and concentration of heavy metals from aqueous solutions (even very dilute ones) by certain types of inactive, dead, microbial biomass (Macaskie *et al.* 1992) Some of the advantages of biosorption include competitive performance, heavy metal selectivity, cost-effectiveness, regenerative and no sludge generation. Sources of biomass include seaweeds, microorganisms (bacteria, fungi, yeast, and molds), activated sludge and fermentation waste. Studies using Biosorbents have shown that both living and dead microbial cell are able to uptake metal ions and offer potential inexpensive alternative to conventional absorbents (Khoo *et al.*, 1991).

However, living cell is subject to toxic effect of the heavy metals, resulting in cell death. Moreover, living cell often require the addition of nutrients and hence increase the biochemical oxygen demand (BOD) and chemical oxygen demand (COD) in the effluent. For these reasons, the use of non-living biomaterials or dead cells as metal binding compounds has been gaining advantage because toxic ions do not affect them. In addition, dead require less care and maintenance, and cheaper (Mofa, 1995). Furthermore, dead biomass could be easily regenerated and reused. The capability of some living microorganisms to accumulate metallic elements has been observed at first from toxicological point of view (Volesky, 1990a, b, c).

However, further researches have revealed that inactive/dead microbial biomass can passively bind metal ions via various physicochemical mechanisms. Therefore researches on biosorption have become an active field for the removal of metal ions or organic compounds. Biosorbent behavior for metallic ions is a function of the chemical make-up of the microbial cells of which it consists (Volesky and Holan, 1995). Mechanisms responsible for biosorption, although understood to a limited extent, may be one or combination of ion exchange, complexation, coordination, adsorption, electrostatic interaction, chelation and micro precipitation (Veglio and Beolchini, 1997; Vijayaraghavan and Yun, 2008; Wang and Chen, 2006). *One of the most important variables affecting the biosorption process is the pH value of the solution (Tsezos, 1990). At lower pH values the active sites of the biomass are protonated and therefore competition between protons and metal ions for the sorption site occurs (Green et al., 1987; Tobin et al., 1984). At low enough pH values, almost all sites become protonated and complete desorption of the bound metal ion is possible (Aldor et al., 1995). Decreasing the pH value to extreme acidic conditions may damage the structure of the biosorbent material. Microscopic observations have shown distorted cells; significant weight loss and decrease in the sorption capacity have been observed (Kuyucak and Volesky, 1989b). With further increase of pH value (>5.5) most heavy metals would precipitate. At those high pH values precipitation may contribute to the overall removal of metals from solution. Furthermore, the pH of a solution can be changed as a result of biosorption. An increase of pH during sorption was reported by Kuyucak and Volesky (1989a), the release of carbonate ions from Ascophyllum biomass could be responsible for the observed increase in solute.*

2.4 Biosorption and Bioaccumulation

Bioaccumulation is defined as the phenomenon of living cells; whereas, biosorption mechanisms are based on the use of dead biomass. Biosorption possesses certain inherent advantages over bioaccumulation processes, which are shown in the Table 2.1.

Table2.1 Compare of biosroption and bioaccumulation.

Feature	Biosroption	Bioaccumulation
Cost	Usually low. Most biosorbents used were industrial, agricultural and other type of waste biomass. Cost involves mainly transportation and other simple processing charges.	Usually high. The process involves living cells and; hence, cell maintenance is cost prone.
PH	The solution pH strongly influences the uptake capacity of biomass. However, the process can be operated under a wide range of pH conditions.	In addition to uptake, the living cells themselves are strongly affected under extreme pH conditions.
Temperature	Since the biomass is inactive, temperature does not influence the process	Temperature severely affects the process.
Maintenance/ storage	Easy to store and use	External metabolic energy is needed for maintenance of the culture
Selectivity	Poor. However, selectivity can be improved by modification /processing of Biomass	Better than biosorption
Versatility	Reasonably good. The binding sites can accommodate a variety of ions.	Not very flexible. Prone to be effected by high metal/salt onditins.
Degree of uptake	Very high. Some biomasses are reported to accommodate an amount of toxicant nearly as high as their dry weight.	Because living cells are sensitive to high toxicant concentration, uptake is usually low.
Rate of uptake	Usually rapid. Most biosorption mechanisms are rapid.	Usually slower than biosorption.Since intracellular accumulation is time consuming.
Toxicant affinity	High under favorable conditions.	Depends on the toxicity of the pollutant.
Regeneration and reuse	High possibility of biosorbent regeneration, with possible reuse over a number of cycles.	Since most toxicants are intracellularly accumulated, the chances are very limited.
Toxicant	With proper selection of elutant, toxicant recovery is possible. In	Even if possible, the biomass cannot be

recovery	many instances, acidic or alkaline solutions proved an efficient medium to recover toxicants.	utilized for next cycle.

2.5 Fixed bed Column

Large-scale application of biosorption in industry is ultimately the main goal for biosorption research. Some metal sequestering biosorbents have been commercialized. For instance, the algal biosorbent, Algasorb, which employed a fresh water algae *Chlorella vulgaris* which employed *Bacillus* biomass were developed for wastewater treatment and metal recovery (Gupta et al., 2000).

The process of metal recovery by biosorbent materials is basically a solid-liquid contact process consisting of the metal uptake cycle and the metal desorption cycle. Appropriate contact between the solution and the solid phase can be accomplished by the fixed packed-bed contactor. The contactor is represented by a column arrangement where the biosorbent is packed into a solid bed, which does not normally move. The liquid usually trickles through the bed in a down flow or up flow arrangement.

Most separation and purification processes that use sorption technology in industry employ continuous flow columns. During this process, the influent is continuously percolated through a sorbent filled column with the undesirable species being retained.

The column sorbent gradually becomes saturated whereby this process begins at the feed zone and gradually progresses to the exit. When the sorbate concentration in the effluent stream reaches a pre-defined level, column operation is terminated. At this point the regeneration process may begin before activation of the next cycle operation (Volesky, 1990).

2.6 Desorption

Biosorption can be used to treat industrial effluents containing heavy metals or to recover precious metals from processing solutions (Davis et al., 2003). The efficiency of biosorption technology depends not only on the efficiency of removal of heavy metals from aqueous solutions but also on the efficiency of regeneration of biosorbent after metals adsorption. The efficiency of desorbing agent or the eluant is often expressed by the S/L ratio, i.e. solid to liquid ratio. The solid represents the solid sorbent (mg dry weight) and the liquid represents the amount of eluant applied (in mL). High values of S/L are desirable for complete elution and to make the process more economical (Gupta et al., 2000). The elution process does not significantly reduce the binding capacity of the biomass and several cycles may be employed. Hydrochloric acid has been recognized as an effective eluant (Brooks, 1991; Holan et al., 1993). Desorption of heavy metals from the biomass is the result of a competition between protons and heavy metal ions under acidic conditions. Active site protonation is favored at low solution pH levels and the competition between protons and metals has been documented by many researchers (Brooks, 1991; Crist et al., 1990; Fourest and Roux, 1992). Generally, increasing the acidity leads to more effective removal of metals from the biomass (Gupta et al., 2000). Desorption process should ensure three major issues: high elution efficiency, low biomass damage and low cost. After the desorption process, the biosorbent is regenerated and a highly concentrated metal solution is obtained. Co-precipitation, flocculation or electro-winning could be used to treat the generated concentrate. Co-precipitation would generate a toxic sludge whereas the solid metal, a more desirable end product, would be recovered from the concentrate by using the electro-winning process (Davis et al., 2003).

2.7 *Mechanism of biosorption*

The mechanisms involved in the process may include ion exchange, coordination, complexation, chelation, adsorption and micro precipitation (Guibal et al., 1992; Fourest and Roux, 1992). These may take place even when cells are metabolically inactive such as when they are killed by physical or chemical methods (Brady et al., 1994). Neither an active membrane transport mechanism nor metabolic energy is required in the process. A non-directed physico-chemical reaction dominates the process (Gadd, 1988 and 1992). Surface and wall binding is a passive process and takes place with both live and dead biomass. This non-metabolic surface binding is very rapid, often less than a few minutes (Khovrychev, 1973). This kind of metal uptake occurs by ion exchange processes involving specific chemical sites on the cell wall (Hancock, 1986).

There are many potential ligands including carboxylate, amine, phosphate, hydroxyl, sulfhydryl, and other functional groups in biomass (Beveridge and Koval, 1981). Sigg (1987) has suggested that amino, phosphate; sulfhydryl, carboxyl or hydroxy groups would be major potential adsorption sites of microbial surface for metal ions. A significant portion of the metal uptake may be due to coordination of the functional groups in the biomass to the metal ions. Tobin et al. (1984) suggested that any site could have several different functional groups participating to various extents in binding the metal ions. Biosorption of heavy metals on different microorganism may involve different functional groups to varying extents (Kapoor and Viraraghavan. 1998a). The polymeric structure of biomass surfaces consists primarily of proteins, carbohydrates, nucleic acids and lipid, and is of a negative charge because of the ionization of organic groups such as carboxylic, aliphatic, aromatic and amino groups and inorganic groups such as hydroxyl and sulphate groups (Hughes and Poole, 1989). Bux and Kasan (1994) suggested that the principal driving force for biosorption of metal ions was the net negative surface charge of the biomass. Therefore, pH will affect the performance of these functional groups.

2.8 Cell wall of algae

Cell walls are more complex in algae than in fungi or bacteria, and three groups of algae, i.e., brown, red and green algae, are of interest, and need to be differentiated in those three evolutionary pathways (Kuyicak and Volesky, 1990; Rincon et al., 2005).

The cell walls of brown algae (Phaeophyta) generally contain three components: cellulose, the structural support; alginic acid, a polymer of mannuronic and guluronic acids (M and G) and the corresponding salts of sodium, potassium, magnesium and calcium; and sulphated polysaccharides (fucoidan matrix). Red algae (Rhodophyta) also contain cellulose, but their interest in connection with biosorption lies in the presence of sulphated polysaccharides made of galactanes (agar and carragenates). Green algae (Chlorophyta) are mainly cellulose, and a high percentage of the cell wall is proteins bonded to polysaccharides to form glycoproteins (Romera et al., 2006).

The Chlorophyta or green algae are extremely varied division. They have chlorophylls a and b along with specific carotenoids, and store carbohydrates as starch. Many of them have cell walls of cellulose. They can present in unicellular, colonial, filamentous, membranous or sheet-like, and tubular types. Green algae are associated with the land plants and have mitochondria with lamellar cristae (Prescott et al., 2002).

Most Rhodophyta or red algae are filamentous and multicellular. The stored food is the carbohydrate called floridean starch. The cell walls of most red algae include a rigid inner part composed of microfibrils and a mucilaginous matrix. The matrix is composed of sulfated polymers of galactose called agar, funori, porphysan, and carrageen an, which are responsible for flexible, slippery texture of the red algae. Agar is used extensively in the laboratory as a culture medium component. Many red algae also deposit calcium carbonate in their cell walls and play an important role in building coral reefs (Prescott *et al.*, 2002).

The Phaeophyta or brown algae have been proved to be most effective biosobent for metal removal, based on statistical review among those algae tested in biosorption (Romera *et al.*, 2006).

Some active sites involved in the metal uptake are determined by using techniques of titration, infra-red (IR) and Raman spectroscopy(RS), electron dispersive spectroscopy (EDS), X-ray photoelectron spectroscopy(XPS), scanning electron microscopy (SEM),nuclear magnetic resonance (NMR), X-ray diffraction analysis (XRD), X-ray absorption fine structure spectroscopy(XAFS) etc. The most important of these groups are summarized by Volesky (2007), including Carbonyl (ketone), Carboxyl, Sulfhydryl (thiol), Sulfonate, Thioether, Amine, Secondary amine, Amide, Imine, Imidazole, Phosphonate , Phosphodiester.

2.9 *Algae and biosorption*

An exhaustive list of algaes used for the uptake of heavy metals as observed in the literature is given in the table (Table 2.2).

Table 2.2 Biosorbent uptake of metals by Algae Biomass

Biomass Type	***Metals***	***Reference***
Spirogyra	*Cu, Pb*	Yi-Chao et al., 2011
Sargassum filipendula	*Cu, Ni*	Kleinübing et al.,2011
Oscillatoria	*Cu , Cd , Pb*	Dhananjay Kumar et al.,2011
Phormidium	*Cu , Cd , Pb*	Dhananjay Kumar et al,. 2011
Sargassum sp	Sm , Pr	Robson C. Oliveiraa et al.2011
Cystoseira stricta	*Pb*	Abdelkader Iddou et al.2011
Sargassum muticum	*Cu*	Roberto Herrero et al.2011
Jania Rubens	*Th*	Cem Gok et al.,2011
Chara globularis	*Pb*	Yan Gao & Xue Yan ,2011
Hydrodictyon reticulatum	*Pb*	Yan Gao & Xue Yan,2011
Sargassum sinicola	*Cd*	Mónica Patron et al., 2011
Sargassum apazeanum	*Cd*	Mónica Patron et al., 2011

Padina australis	Ni	Pahlavanzadeh et al. 2010
Spirulina platensis	Cr	Elisabetta Finocchio et al., 2010
Sargassum sinicola	Cu , Cd	Monica Patron et al., 2010
Turbinaria Conoides	La, Ce, Eu, , Yb	Vijayaraghavan et al., 2010
Sargassum glaucescens	Ni	Pahlavanzadeh et al., 2010
Cladophora hutchinsiae	Se	Mustafa Tuzen et al., 2010
Hypnea valentiae	Cd	Aravindhan Rathinam et al.2010
Desmodesmus pleiomorphus	Cd	Cristina M. et al., 2010
Posidonia oceanica	Cd	Marta Izquierdo et al., 2010
Oedogonium hatei	Ni	Gupta et al., 2010
Geobacillus hermodenitrificans	Cr, Cu, Zn ,Cd ,Ag ,Pb	Chatterjee et al., 2010
Isochrysis galbana	Cu, Zn ,Ni , Co , Cd, Pb	Elena Lourie et al., 2010
Nanochloropsis oceancia	Cu, Zn ,Ni , Co , Cd, Pb	Elena Lourie et al., 2010
Chlorella sp.	Cu, Zn ,Ni , Co , Cd, Pb	Elena Lourie et al., 2010
Pavlova lutheri	Cu, Zn ,Ni , Co , Cd, Pb	Elena Lourie et al.,2010
Pseudokirchneriella	Cu, Zn ,Ni , Co , Cd, Pb	Elena Lourie et al., 2010
Sargassum filipendula	Ni	Marcia Teresinha et al., 2009
Scenedesmus obliquus	Cd	Cristina M. et al. ,2009
Chaetomorpha linum	Cu , Zn	Leila Chebil Ajjabi et al., 2009
Sargassum filipendula	U	Rodrigues Silva et al., 2009
Turbinaria conoides	Rb	Siew-Ling Hii,2009
Lobophora Wariegata	Cd , Pb	Shaik Basha et al., 2008
Oscillatoria sp.	Cd	Hikmet Katırcıog et al.,2008
Spirulina platensis	Pd , Cd , Ni	Aysegul et al., 2008
Pithophora oedogonia	Cu , Pb	Alpana Singh et al. ,2008
Spirulina platensis	U	Arun Dheetcha et al., 2008
Sargassum sp.	Pb, Cu, Cd	Ping Xin Sheng et al. ,2007
Spirulina	Cr , Cu , Ni	Hiren Doshi et al. ,2007
f Chlorella sp.	Cu , Ni	Hiren Doshi et al. ,2006
Ecklonia sp.	Cr, Zn	Donghee Park et al. ,2006
Sargassum sp.	Cu	Paul Chen et al., 2006

Sargassum sp.	Cu, Cd, Zn, Ni	Paul Chen et al., 2006
Spirulina sp.	Cr, Cd, Cu	Chojnacka et al., 2005
Ulva reticulata	Co, Cu, Ni	Vijayaraghavan et al., 2005
Chlamydomonas reinhardtii	Hg, Pb, Cd	Tuzun et al., 2005
Cystoseira baccata	Hg	Herrero et al., 2005
Padina sp.	Cu, Ni	Sheng et al., 2004
Chlorella vulgaris	Cu , Ni	Nirupama Mallick,2003
Sargassa sp.	Zn	Davis et al., 2003
Gloeothece magna	Cd	Zakaria A. Mohamed, 2001
Spirogyra sp.	Cr	Gupta et al., 2001
Chlorella vulgaris	Cd, Pb, Ni, Zn	Klimmekand et al., 2001
Arthrobacter sp.	Cu	Beolchini et al., 2001
Arthrobacter sp.	Cu, Cd, Fe	Pagnanelli et al., 2000
Sargassum	Fe	Figuria et al., 1999
Arthrobacter sp.	Cu	Veglio et al. 1998
Sargassum natans	Ag , Ni, Pb	Volesky and Kuyucak, 1988
Palmaria tevera	Au	Kuyuack and Volesky, 1988
Chlorella pyrenoidosa	Au	Darnall et al., 1988
Chlorella vulgaris	Au	Gee and Dudeney, 1988
Chondrus crispus	Au, Co	Kuyuack and Volesky, 1988
Spirulina platensis	Au	Darnall et al., 1988
Ascophyllum nodosum	Au	Kuyuack and Volesky, 1988
Sargassum fluitans	Fe	Figueira et al., 1995
Fucus vesiculosus	Ni, Pb	Holan and Volesky, 1994
Ascophylum nodosum	Ni, Pb	Holan and Volesky, 1994
Ascophyllum nodosum	Cd	Holan et al., 1993
Sargassum natans	Cd	Holan et al., 1993
Fucus vesiculosus	Cd	Holan et al., 1993
Scenedesmus obliquus	Cu	Mattuschka et al., 1993

Thus for, the literature survey suggests that the work aiming to study applicability of biomass especially algae to remove heavy metal contamination by automobile industry has not been carried out. It is in this context the present work may be considered as some addition to the knowledge of this kind. It may not be and overstatement if the articles based of the present study are published in the international journals.

Chapter III Materials and Methods

3.1 Cultivation of algaes

In this project we collected algae from automobile industrial effluent and after identify of them, we understood three types of algae were on the effluent treatment plant that includes the *Oscillatoria sp., Chlorella Pyrenoidosa* and *Spirulina sp.* *Chlorella pyrenoidosa* (2738) and *spirulina sp.* (5143) obtained from National Collection of

Industrial Microorganisms (NCIM) from pune and oscillatoria sp. was obtaining from agriculture collage from pune, which was isolated and thoroughly pure.

The *oscillatoria sp.*, *spirulina sp.* and *Chlorella pyrenoidosa* were maintained in modified Bold' basal medium (MBBM), Spirulina Media and Fog's Media at 28 ^{0}C with using 3000 lx light intensity respectively. After 21 days cultivation period cells were harvested by centrifugation and were washed several times with deionised Water in order to remove culture media and were kept on a filter paper to reduce the water content. The biomasses dried at 60 °C in an oven for 24 h and milled to a gritty consistency. The biomasses were sieve for particle size smaller than 1 mm and stored in dark bottle and keeps in a dry cabinet for experiments. All of the media are C and 15 lbs for 20 min. sterilized by autoclaving at 121

- **Fog's Medium Formula:** MgSO4.7H2O 0.2 g ,K2HPO4 0.2 g,*Micronutrient 1.0 ml solutionCaCl2.H2O 0.1 g ,*Fe-EDTA solution 5.0 ml , Distilled water 1.0 l , Agar(Difco) 12.0 g Adjust pH to 7.5

*Micronutrient solution : H3BO3 286.0 mg ,MnCl2.4H2O 181.0 mg,ZnSO4.7H2O 22.0 mg,Na2MoO4.2H2O 39.0 mg,CuSO4.5H2O 8.0 mg, Distilled water 100.0 ml

*Fe-EDTA solution

In hot water dissolve 745.0 mg of Na2EDTA and then add 557.0 mg of FeSO4.7H2O. Boil the solution for few minutes and make the volume to 100.0 ml.

- **Spirulina Media Formula:**NaHCO3 10.0 g, $NaNO_3$ 2.5 g, NaCl 1.0 g , K_2HPO_4 0.5 g ,K_2SO_4 1.0 g ,$MgSO_4.7H_2O$ 0.2 g ,$CaCl_2$ 0.04 g , $FeSO_4$.7H2O 0.01 g ,Distilled water 1.0 L ,Agar (Difco) 20.0 g ,
- **MBBM Formula :** For preparation MBBM to 950 mL of d-H_2O add:10.0 mL of stock solutions 1, 2, 3, 4, 5 and 1.0 mL of stock solutions 7, 8 and 2.0 mL of stock solution 10 ,1.0 gm of bacto-peptone and 5.0 gm of sucrose.1) 25.0 gm $NaNO_3$ 2) 2.5 gm $CaCl_2.2H_2O$ 3) 7.5 gm $MgSO_4$.7H2O 4) 7.5 gm K_2HPO_4 5) 17.5 gm KH_2PO_4 6) 2.5 gm NaCl 7) 50.0 gm disodium EDTA, 31.0 gm KOH 8) 4.98 gm FeSO4.7H_2O 9) 11.42 gm H_3BO_3 10) 8.82 gm $ZnSO_4.7H_2O$, 1.44 gm

$MnCl_2.4H_2O$, 0.71 gm MoO_3, 1.57 gm $CuSO_4.5H_2O$, and0.49 gm $CoNO_3.6H_2O$ per liter d-H_2O.MBBM.

3.2 Pretreatment of biomass

In the pretreatment, the biomass was slowly stirred in the chemical solution for a certain period of time .In this study $CaCl_2$ has the best result with in chemical solution. After each pretreatment the biomass was washed with generous amounts of double distilled water and then dried in an oven at 60 °C for 24 hours.

3.3 Preparation of glass wares and stock solution samples

Cleaning of glassware used in the experiments was done as follows: First, washed with detergent solution, then rinsed with tap water; secondly, rinsed with 10% nitric acid, and then rinsed with tap water; finally rinsed with distilled water. The cleaned glassware was dried prior to use in experiments. Samples of automobile industrial effluent collected on TATA MOTORS LIMITED from Pune- INDIA that filtrated by whatman paper number 0.45 and after analysis of samples we understood four type heavy metals were on the effluent that include: Chromium, Copper, Nickel and Zinc. For the laboratory study a stock solution of 1000 mg/l of Cr (VI) , Cu(II) , Ni(II) , Zn(II) were obtained by dissolving potassium dichromate , copper sulphate, nickel chloride and zinc chloride (Merck Company) in distilled water respectively. The test solutions of various concentrations range from 10 mg/L to 100 mg/L were prepared from the stock solution. The solution pH was adjust using 0.1 M HNO_3 and 0.1 M NaOH at the beginning of the experiment and not controlled afterwards. The conical flasks (250 mL) were shaken at 120 rpm in a temperature controlled rotatory shaker. Ions were determined by atomic absorption spectrophotometer (**UNICAM, Model 929, UK**).

3.4 Batch biosorption studies

Batch mode adsorption studies for individual metal compounds were carried out to investigate the effect of different parameters such as adsorbate concentration, adsorbent dose, agitation time and pH. Solution containing adsorbate and adsorbent was taken in 250 mL capacity conical flask and agitated at 120 rpm in a shaker at predetermined time intervals. The adsorbate was decanted and separated from the adsorbent using Whatman No.41 filter paper (Vijayaraghavan *et al.* 2008).

3.5 Effect of contact time

For the determination of rate of metal biosorption by biomasses from 100 ml (at 10, 20, 50, 100 mg/L) on conical flask 250 mL, the supernatant was analyzed for residual metal at different time intervals. The pH and the adsorbent dosage was kept constant at pH= 5± 0.01, which varied according to the adsorbent and adsorbate under consideration. Amount of biomass dosage was 0.1±0.001 g for all three biomass (*Chlorella pyrenoidosa, Oscillatoria sp.* and *Spirulina sp.*) and temperature was 25 ±1 °C and agitation speed of shaker was 120 rpm.

3.6 Effect of pH

To determine the effect of pH on the adsorption of metal solutions (100 mL) of different concentration ranges (10, 20 and 50 mg/L)) at conical flask 250 mL were adjusted to desired pH values and mixed with constant amount of adsorbent and agitated at preset equilibrium time. The equilibrium time and adsorbent dosage varied with the metal and adsorbent under consideration. Amount of biomass dosage was 0.1±0.001 g for all three biomass (*Chlorella pyrenoidosa, Oscillatoria sp.* and *Spirulina sp.*) and temperature was 25 ±1 °C and agitation speed of shaker was 120 rpm and contact time was 120 min.

3.7 Effect of adsorbent dosage

The effect of adsorbent dosage i.e., the amount of the biomasses on the adsorption of metals was studied at different dosages ranging from 0.1 to 3 g with varied metal concentrations of 10, 20 and 50 mg/L. The equilibrium time and the pH were kept constant depending on the metal under consideration. The pH and the adsorbent dosage was kept constant at pH= 5± 0.01, which varied according to the adsorbent and adsorbate under consideration. Agitation time was 120 minute for all three biomass (*Chlorella pyrenoidosa, Oscillatoria sp.* and *Spirulina sp.*) and temperature was 25 ±1 °C and agitation speed of shaker was 120 rpm.

3.8 Effect of temperature

Optimum biomass concentration with optimum pH was used to monitor the temperature effect on biosorption. Experiments were carried out at different temperatures from 10-40 °C for each culture and kept on rotary shaker at 120 rpm. The samples were allowed to attain equilibrium. To determine the effect of temperature on the adsorption of metal solutions (100 mL) of concentration 50 mg/L at conical flask 250 mL were adjusted to desired pH values and mixed with constant amount of adsorbent and agitated at preset equilibrium time. The equilibrium time and adsorbent dosage varied with the metal and adsorbent under consideration. Amount of biomass dosage was 0.1±0.001 g for all three biomass (*Chlorella pyrenoidosa, Oscillatoria sp.* and *Spirulina sp.*) and pH was 5±0.001 and agitation speed of shaker was 120 rpm and contact time was 120 min.

3.9 Desorption studies

After adsorption, the adsorbates – loaded adsorbent were separated from the solution by centrifugation and the supernatant was drained out. The adsorbent was gently washed with water to remove any unadsorbed adsorbate. Regeneration of adsorbate from the adsorbate – laden adsorbent was carried out using the desorbing media – distilled water at pH ranges using dilute solutions of EDTA, HCL and HNO_3 (Stirred at 120 rpm for 120 min at 25 ^{0}C) . Then they were agitated for the equilibrium time of respective adsorbate. The desorbed adsorbate in the solution was separated and analyzed for the residual heavy metals.

3.10 Equilibrium isotherms

The isotherm studies were performed in the solution with the initial concentrations ranging from 10 to 100 mg/L at optimum pH values for ions (pH=4.5 or pH = 5) .After shaking the flask containing the mixture of biomass (120 rpm, 25 °C) and ions for 120 min, the amount of residual ions in the filtrated solution was analyzed. The biosorption equilibrium uptake capacity for each sample was calculated according to mass balance on the ions expressed in this equation: $q_e = \frac{(C_0 - C_e)}{M} \times V$

where V is the sample volume (L), C_0 is the initial ion concentration (mg/L), C_e is the equilibrium or final ion concentration (mg/L) , M is the biomass dry weight (g), and q_e is the biomass biosorption equilibrium ions uptake capacity (mg/g).Langmuir and Freundlich isotherms, the two classical adsorption models, were used to describe the equilibrium between adsorbed ions on the biomass cell (q_e ,q) and ions in the solution (C_e, q) in this study (Chen and Li 2010; Allen et al. 2004).

Langmuir isotherm model (Langmuir I. 1918): $$q_e = \frac{q_{max} C_e b}{1 + C_e b}$$

That after arrange we have; $$\frac{C_e}{q_e} = \frac{1}{q_{max} b} + \frac{C_e}{q_{max}}$$

These values q_{max} and b can be obtained from the slopes and the intercepts of the linear plots respectively, where experimental data of Ce/qe as the function of C_e. .

The empirical Freundlich equation based on sorption on a heterogeneous surface, on the other hand, is as follows (Freundlich H. 1906): $q_e = K_f (C_e)^n$

The equation can be linearized in the following logarithmic form:

$$\ln q_e = \ln k_f + \frac{1}{n} \ln C_e$$

These values n & K_f can be obtained from the slopes and the intercepts of the linear plots respectively, where experimental data of Ln q_e as the function of Ln C_e. *.The Langmuir and* Freundlich *models have been used successfully to describe equilibrium biosorption*.The value of n is more than 1, which represent favorable adsorption condition (McKay *et al.* 1982).

3.11Equilibrium parameter R_L

The essential characteristics of a Langmuir isotherm can be expressed in terms of a dimensionless constant separation factor or equilibrium parameter R_L, which is defined by $R_L = 1/1+bC_o$

C_o is the initial adsorbate concentration (mg/L) and b is the Langmuir constant (L/mg). The parameter indicates the shape of the isotherm as follows (Table3.1):

Table 3.1: Type of isotherm for various R_L

R_L	$R_L>1$	$R_L=1$	$0< R_L<1$	$R_L = 0$
Type of isotherm	Un favorable	Linear	Favorable	Irreversible

The R_L values at different initial adsorbate concentrations indicate favorable adsorption for all the adsorbents and adsorbates studied.

3.12 Kinetic modeling

The experimental break through curves for the effects of contact time on a bound rate of ions. Therefore, for the following experiments, the contact time was maintained for 60 min to ensure that equilibrium was fully achieved. The pseudo-second-order equation is also based on the sorption capacity, which is expressed as:

$$\frac{t}{q_t} = \frac{1}{\left(K_2 q^2{}_e\right)} + \frac{t}{q_e}$$

K_2 is the rate constant of pseudo-second-order sorption (g· mg^{-1}·min^{-1}). K_2qe^2 is the initial rate constant (represented by h, mg·g^{-1}·min^{-1}). Plotting t/q_t versus t will give a straight line. The values of q_e and K_2 can be determined from the slope and intercept of the plot, respectively. The results showed that the pseudo-second-order model fitted the simulation curve much better than the pseudo-first-order model for ions (Ho *et al* 1996).

3.13 Fixed bed column

3 g of dry biomass were packed into a 50 cm long column of 2.5 cm inner diameter made of stainless still. Glass bead was placed at the bottom and up of the column to ensure homogenous distribution of the feeding solution. *The biomass was first washed by slowly flooding the column with distilled water from the up. After washing the biomass with distilled water and prior to the sorption of heavy metals the biomass was washed with a 0.1 M $CaCl_2$ solution in order to prepare the Ca-biomass. The pH value in the storage tank was adjusted to 5.0 using a 0.1M HCL and NaOH solution. The metal solutions were fed into the column from the up, at a flow rate equal to 1.67ml/min allowing 15 and 30 minutes of detention time in the column. Samples of column effluent were collected at preset time intervals and were analyzed for heavy metals content using AAS. When the biomass in the column was saturated with metal, that is, the final column effluent concentration approached the influent* *concentration; the column feed was switched to distilled water for several hours and then followed by a 0.1 M HNO_3 solution in order to elute the heavy metals. For the operation of a Cr, Cu, Ni and Zn biosorption column, an influent, pH= 5.0 and an initial concentration of 10 mg.L^{-1} of heavy metals, with an upward flow rate of 1.67 mL/min was used. The column was loaded with 3 g (dry weight) of biomass.*

3.14 Thermodynamic studies:

Adsorption, studies in the temperature range 283–313 K were conducted to determine thermodynamics constants such as Gibbs free energy change (ΔG°), enthalpy change (ΔH°) and entropy change (ΔS°) for the system and to ascertain the sorption mechanism. For this study, adsorbent dosage selected was 0.1 gr. and chromium ,copper , nickel and zinc concentration was 50 mg L^{-1} with pH= 5 in a conical flask and allowed to equilibrate for 1 h at the different temperatures ranging from 283 to 313 K.

In order to determine thermodynamic parameters, experiments were carried out at different temperatures in the range of 283–313 K for heavy metals adsorption. The thermodynamic parameters such as standard Gibbs free energy change (ΔG°), enthalpy change (ΔH°) and entropy change (ΔS°) were estimated to evaluate the feasibility and nature of the adsorption process. The Gibbs free energy change, of the process is related to equilibrium constant by the equation:

$$\Delta G^{\circ} = -RT \ln K_C$$

where, T is temperature in K, R is ideal gas constant having value as 8.314, J mol^{-1} K^{-1} and K_C is thermodynamic equilibrium constant. The thermodynamic equilibrium constant, was determined as:

$$K_C = \frac{C_a}{C_e}$$

Where, C_a is mg of adsorbent adsorbed per liter and C_e is the equilibrium concentration of solution, mg/L.

According to thermodynamics, the Gibbs free energy change is also related to the enthalpy change (ΔH°) and entropy change (ΔS°) at constant temperature by the following Van' t Hoff equation:

$$\ln K_C = \frac{-\Delta H^\circ}{R}\frac{1}{T} + \frac{\Delta S^\circ}{R}$$

The values of enthalpy change (ΔH°) and entropy change (ΔS°) were calculated from the slope and intercept of the plot Ln Kc versus, 1/T (Bansal *et al* 1982)

3.15 FT-IR spectroscopy (Fourier Transform Infrared)

In order to determine the functional groups responsible for heavy metals (Cr, Cu, Ni and Zn) biosorption, IR spectroscopy was used that about 0.1 g biomass was mixed with KBr for FT-IR spectra analysis (**Shimadzu, Model 8400**).

3.16 SEM (Scanning Electron Microscopy)

The SEM was used to investigate the morphology of the biosorbent. We used samples with pH=5 and C_0 = 10 mg.L^{-1}. Scanning Electron Microscope (**SEM, JEOL, JSM-6360A**) used for this study.

Chapter IV: Result and Discussion

4.1 Effect of pH for Chromium

Results on the effect of pH of Cr at different initial metal ion concentrations by *Chlorella pyrenoidosa, Oscillatoria sp.* and *Spirulina sp.* present in the Figures 4.1 to 4.6.

Maximum percentage of biosorption accursed at the initial concentration of 10 mg/L at the time of 120 minute at pH= 5 for Cr was 92.51% , 97.92% and 82.67% and metal ion uptake capacity was 9.25 mg/g , 9.79 mg/g and 8.26 mg/g and when initial concentration of Cr increased to 50 mg/L , percentage of biosortion of Cr was 89.66 % , 69.12 % and 63.44 % and uptake capacity was 41.82 mg/g , 22.41 mg/g and 34.56 mg/g for *Chlorella pyrenoidosa, Oscillatoria sp.* and *Spirulina sp.* respectively.

With increase the initial concentration percentage of biosorption decreased and metal ion uptake capacity was increased.In effect of pH and initial concentration of 10 mg/L, percentage of remove of Cr and metal ion uptake capacity *for the three biomasses under study has been established as:*

Oscillatoria sp. > Chlorella pyrenoidosa > Spirulina sp.

when initial concentration increased to 50 mg/L, percentage of remove of Cr and metal ion uptake capacity *for the three biomasses under study has been established as: Chlorella pyrenoidosa > Oscillatoria sp. > Spirulina sp.*

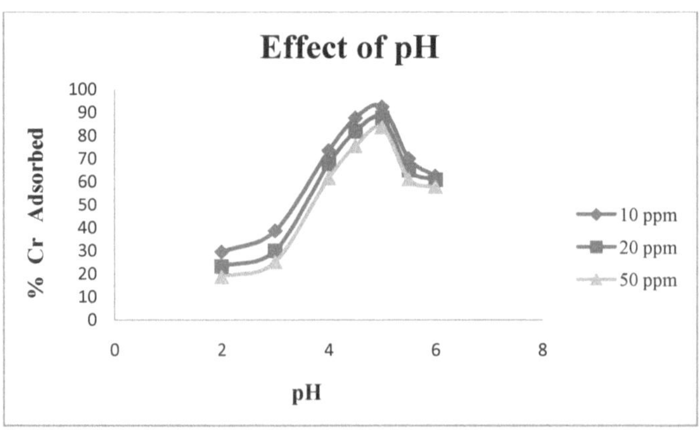

Fig4.1 Effect of pH on biosorption of Cr by *Chlorella pyrenoidosa* (biomass dose=0.1 g, initial Cr ion concentration = 10, 20 and 50 mg/L; temperature = 25 ˚C; agitation speed = 120 rpm; contact time = 120 min).

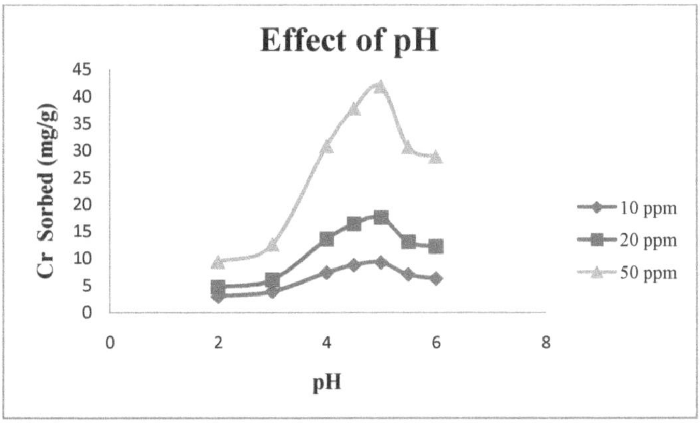

Fig4.2 Effect of pH on biosorption of Cr by *Chlorella pyrenoidosa* (biomass dose=0.1 g, initial Cr ion concentration = 10, 20 and 50 mg/L; temperature = 25 ˚C; agitation speed = 120 rpm; contact time = 120 min).

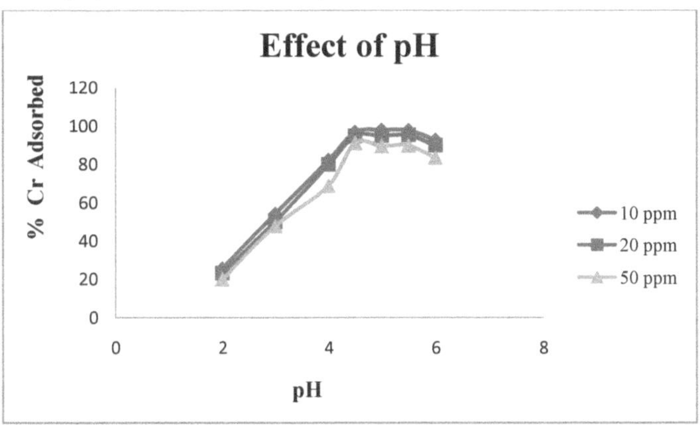

Fig4.3 Effect of pH on biosorption of Cr by *Oscillatoria sp.* (Biomass dose=0.1 g, initial Cr ion concentration = 10, 20 and 50 mg/L; temperature = 25 °C; agitation speed = 120 rpm; contact time = 120 min).

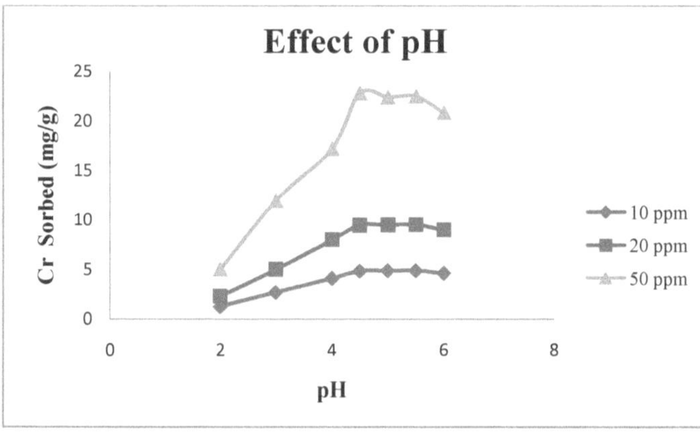

Fig4.4 Effect of pH on biosorption of Cr by *Oscillatoria sp.* (Biomass dose=0.1 g, initial Cr ion concentration = 10, 20 and 50 mg/L; temperature = 25 °C; agitation speed = 120 rpm; contact time = 120 min).

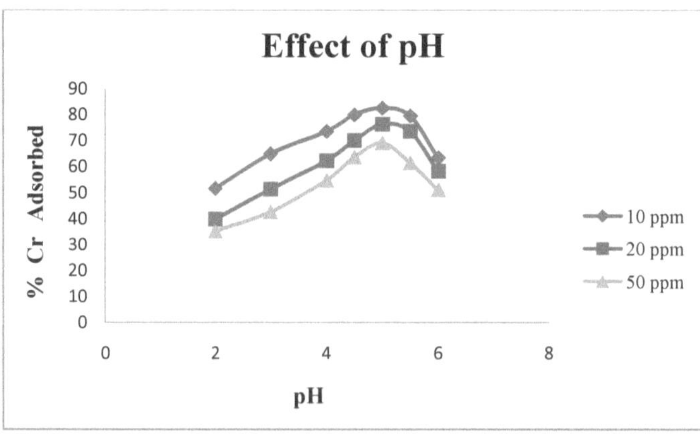

Fig.4.5 Effect of pH on biosorption of Cr by *Spirulina sp.* (Biomass dose=0.1 g, initial Cr ion concentration = 10, 20 and 50 mg/L; temperature = 25 °C; agitation speed = 120 rpm; contact time = 120 min).

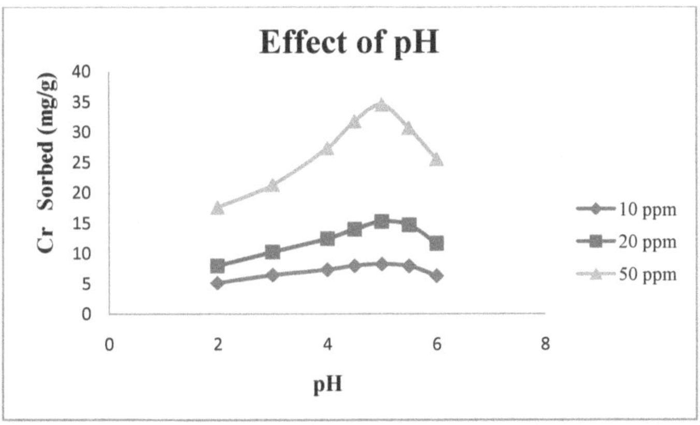

Fig4.6 Effect of pH on biosorption of Cr by *Spirulina sp.* (Biomass dose=0.1 g, initial Cr ion concentration = 10, 20 and 50 mg/L; temperature = 25 °C; agitation speed = 120 rpm; contact time = 120 min).

4.2 Effect of pH for Copper

Results on the effect of pH of Cu at different initial metal ion concentrations by *Chlorella pyrenoidosa, Oscillatoria sp.* and *Spirulina sp.* present in the Figures 4.7 to 4.12.

Maximum percentage of biosorption accursed at the initial concentration of 10 mg/L at the time of 120 minute at pH= 5 for Cu was 88.48 % , 92.08 % and 90.09 %and metal ion uptake capacity was 8.85 mg/g , 9.21 mg/g and 9.01 mg/g and when initial concentration of Cu increased to 50 mg/L , percentage of biosortion of Cu was 80.13

%,90.92 % and 84.44% and uptake capacity was 40.06 mg/g , 45.46 mg/g and 42.22 mg/g for *Chlorella pyrenoidosa, Oscillatoria sp.* and *Spirulina sp.* respectively.

With increase the initial concentration percentage of biosorption decreased and metal ion uptake capacity was increased. In effect of pH and initial concentration of 10 mg/L, percentage of remove of Cu and metal ion uptake capacity *for the three biomasses under study has been established as:*

Oscillatoria sp. > Spirulina sp. > Chlorella pyrenoidosa

when initial concentration increased to 50 mg/L, percentage of remove of Cu and metal ion uptake capacity *for the three biomasses under study has been established as: Oscillatoria sp. > Spirulina sp. > Chlorella pyrenoidosa*

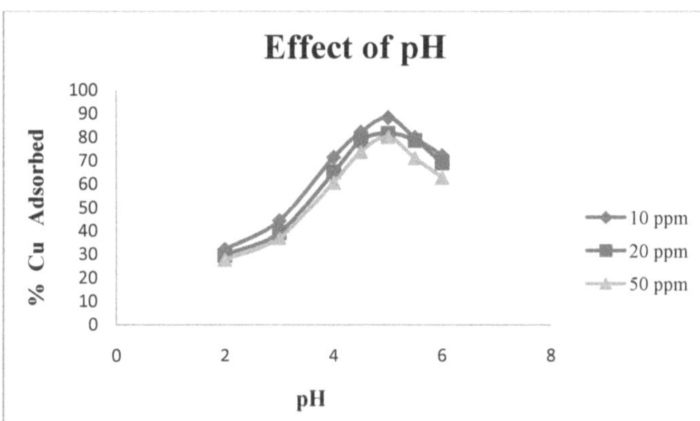

Fig4.7 Effect of pH on biosorption of Cu by *Chlorella pyrenoidosa* (biomass dose=0.1 g, initial Cu ion concentration = 10, 20 and 50 mg/L; temperature = 25 °C; agitation speed = 120 rpm; contact time = 120 min).

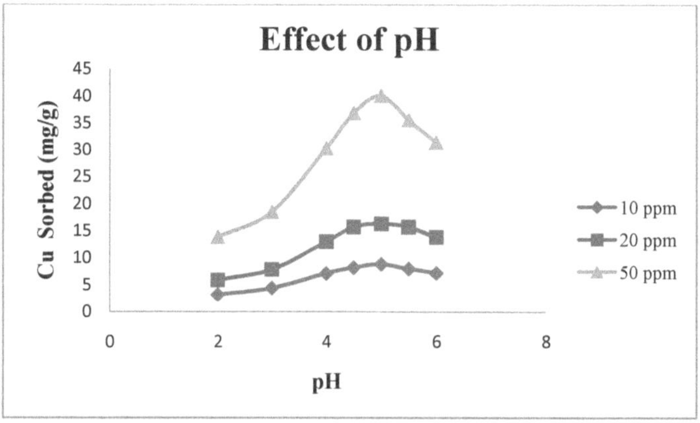

Fig4.8 Effect of pH on biosorption of Cu by *Chlorella pyrenoidosa* (biomass dose=0.1 g, initial Cu ion concentration = 10, 20 and 50 mg/L; temperature = 25 ˚C; agitation speed = 120 rpm; contact time = 120 min).

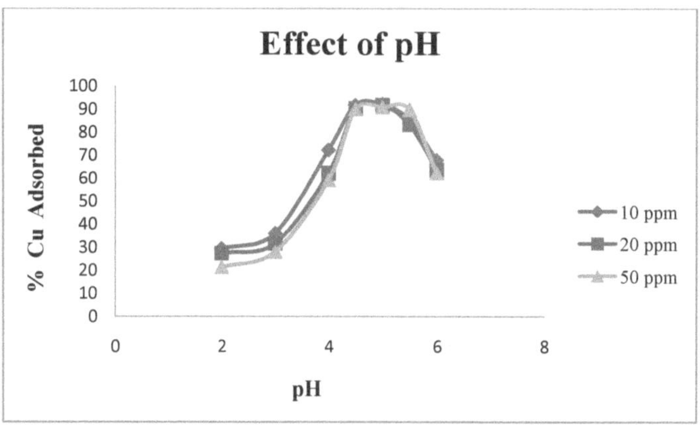

Fig.4.9 Effect of pH on biosorption of Cu by *Oscillatoria sp.* (biomass dose=0.1 g, initial Cu ion concentration = 10, 20 and 50 mg/L; temperature = 25 ˚C; agitation speed = 120 rpm; contact time = 120 min).

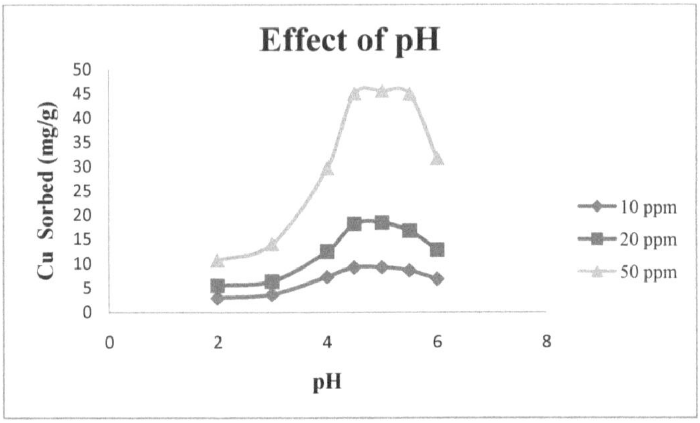

Fig.4.10 Effect of pH on biosorption of Cu by *Oscillatoria sp.* (biomass dose=0.1 g, initial Cu ion concentration = 10, 20 and 50 mg/L; temperature = 25 °C; agitation speed = 120 rpm; contact time = 120 min).

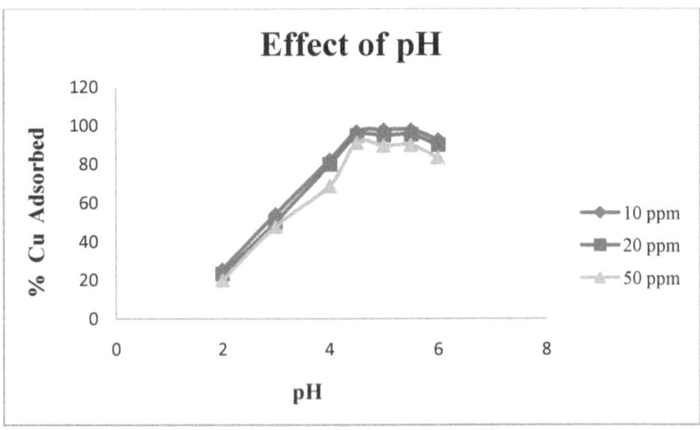

Fig4.11 Effect of pH on biosorption of Cu by *Spirulina sp.* (biomass dose=0.1 g, initial Cu ion concentration = 10, 20 and 50 mg/L; temperature = 25 °C; agitation speed = 120 rpm; contact time = 120 min).

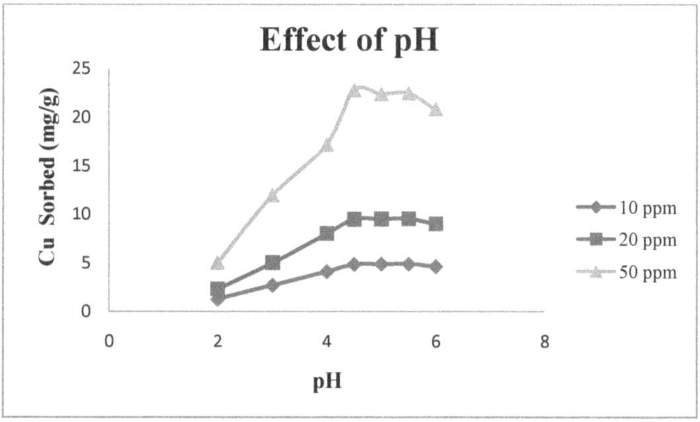

Fig4.12 Effect of pH on biosorption of Cu by *Spirulina sp.* (biomass dose=0.1 g, initial Cu ion concentration = 10, 20 and 50 mg/L; temperature = 25 °C; agitation speed = 120 rpm; contact time = 120 min).

4.3Effect of pH for Nickel

Results on the effect of pH of Ni at different initial metal ion concentrations by *Chlorella pyrenoidosa, Oscillatoria sp.* and *Spirulina sp.* present in the Figures 4.13 to 4.18.

Maximum percentage of biosorption accursed at the initial concentration of 10 mg/L at the time of 120 minute at pH= 5 for Ni was 94.31% , 82.31 % and 83.60 % and metal ion uptake capacity was 9.43 mg/g , 8.23 mg/g and 8.36 mg/g and when initial concentration of Ni increased to 50 mg/L , percentage of biosortion of Ni was 86.60 %,70.63 % and 74.05 % and uptake capacity was 43.30 mg/g , 35.31 mg/g and 37.02 mg/g for *Chlorella pyrenoidosa, Oscillatoria sp.* and *Spirulina sp.* respectively.

With increase the initial concentration percentage of biosorption decreased and metal ion uptake capacity was increased.In effect of pH and initial concentration of 10 mg/L, percentage of remove of Ni and metal ion uptake capacity *for the three biomasses under study has been established as:*

Chlorella pyrenoidosa > Spirulina sp. > Oscillatoria sp.

when initial concentration increased to 50 mg/L, percentage of remove of Ni and metal ion uptake capacity *for the three biomasses under study has been established as:*

Chlorella pyrenoidosa > Spirulina sp. > Oscillatoria sp.

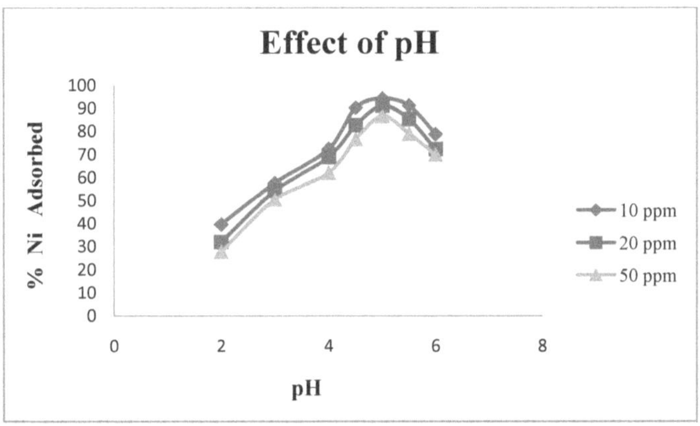

Fig4.13 Effect of pH on biosorption of Ni by *Chlorella pyrenoidosa* (biomass dose=0.1 g, initial Ni ion concentration = 10, 20 and 50 mg/L; temperature = 25 °C; agitation speed = 120 rpm; contact time = 120 min).

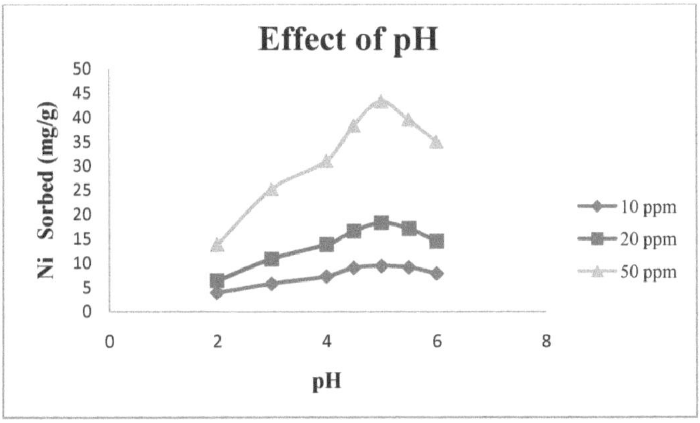

Fig4.14 Effect of pH on biosorption of Ni by *Chlorella pyrenoidosa* (biomass dose=0.1 g, initial Ni ion concentration = 10, 20 and 50 mg/L; temperature = 25 ˚C; agitation speed = 120 rpm; contact time = 120 min).

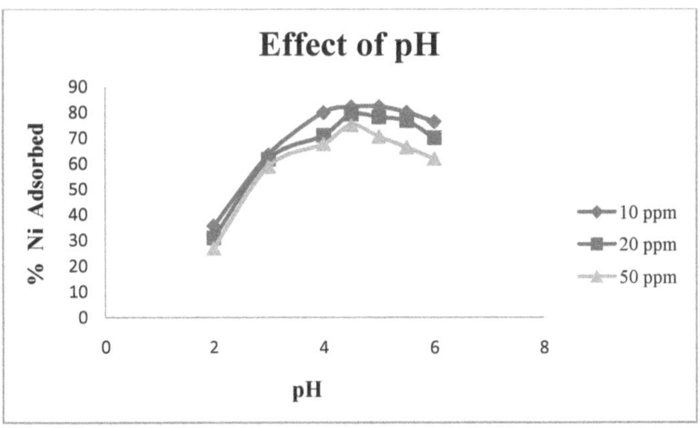

Fig4.15 Effect of pH on biosorption of Ni by *Oscillatoria sp.* (biomass dose=0.1 g, initial Ni ion concentration = 10, 20 and 50 mg/L; temperature = 25 ˚C; agitation speed = 120 rpm; contact time = 120 min).

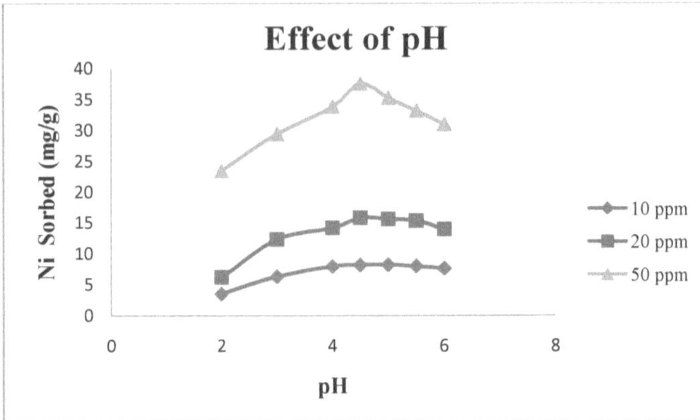

Fig4.16 Effect of pH on biosorption of Ni by *Oscillatoria sp.* (biomass dose=0.1 g, initial Ni ion concentration = 10, 20 and 50 mg/L; temperature = 25 °C; agitation speed = 120 rpm; contact time = 120 min).

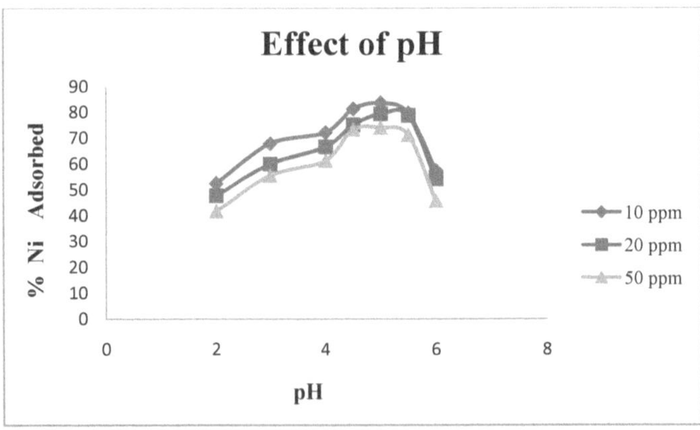

Fig4.17 Effect of pH on biosorption of Ni by *Spirulina sp.* (biomass dose=0.1 g, initial Ni ion concentration = 10, 20 and 50 mg/L; temperature = 25 °C; agitation speed = 120 rpm; contact time = 120 min).

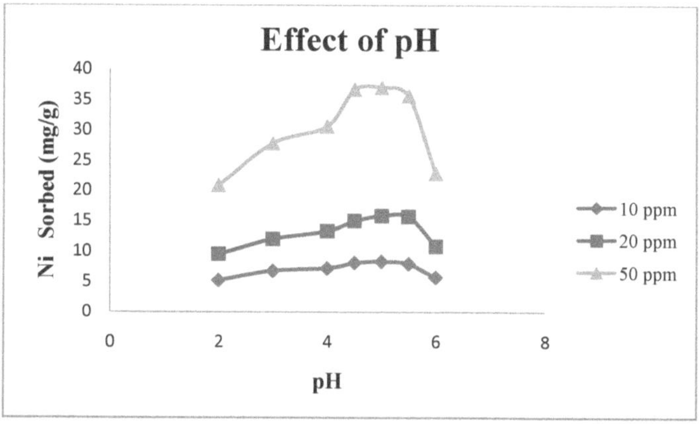

Fig4.18 Effect of pH on biosorption of Ni by *Spirulina sp.* (biomass dose=0.1 g, initial Ni ion concentration = 10, 20 and 50 mg/L; temperature = 25 ˚C; agitation speed = 120 rpm; contact time = 120 min).

4.4 Effect of pH for Zinc

Results on the effect of pH of Zn at different initial metal ion concentrations by *Chlorella pyrenoidosa pyrenoidosa, Oscillatoria sp.* and *Spirulina sp.* present in the Figures 4.19 to 4.24.

Maximum percentage of biosorption accursed at the initial concentration of 10 mg/L at the time of 120 minute at pH= 5 for Zn was 85.55% , 83.07 %and 86.69% and metal ion uptake capacity was 8.55 mg/g , 8.31 mg/g and 8.67 mg/g and when initial concentration of Zn increased to 50 mg/L , percentage of biosortion of Zn was 71.33 % , 72.20% and 72.23% and uptake capacity was 35.66 mg/g , 46.10 mg/g and 36.11 mg/g for *Chlorella pyrenoidosa, Oscillatoria sp.* and *Spirulina sp.* respectively.

With increase the initial concentration percentage of biosorption decreased and metal ion uptake capacity was increased.In effect of pH and initial concentration of 10 mg/L, percentage of remove of Zn and metal ion uptake capacity *for the three biomasses under study has been established as:*

Spirulina sp. > Chlorella pyrenoidosa > Oscillatoria sp.

when initial concentration increased to 50 mg/L, percentage of remove of Zn and metal ion uptake capacity *for the three biomasses under study has been established as: Spirulina sp. > Oscillatoria sp. > Chlorella pyrenoidosa*

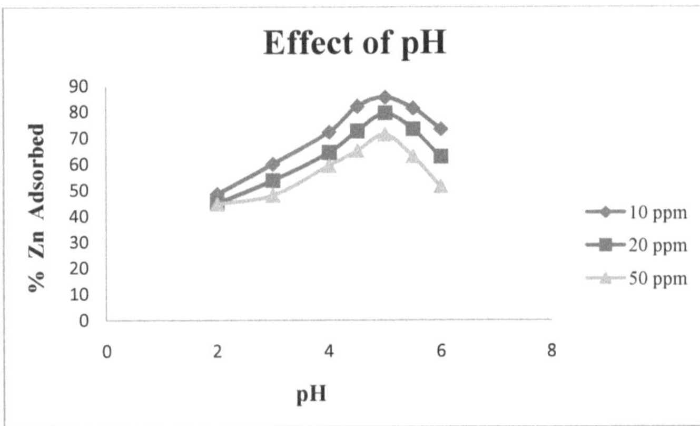

Fig4.19 Effect of pH on biosorption of Zn by *Chlorella pyrenoidosa* (biomass dose=0.1 g, initial Zn ion concentration = 10, 20 and 50 mg/L; temperature = 25 °C; agitation speed = 120 rpm; contact time = 120 min).

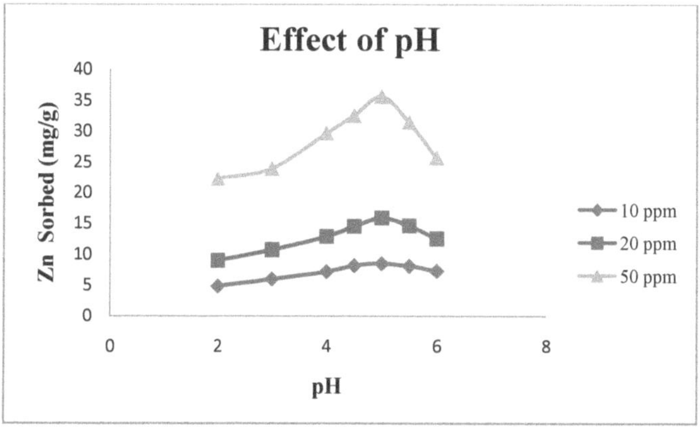

Fig4.20 Effect of pH on biosorption of Zn by *Chlorella pyrenoidosa* (biomass dose=0.1 g, initial Zn ion concentration = 10, 20 and 50 mg/L; temperature = 25 ˚C; agitation speed = 120 rpm; contact time = 120 min).

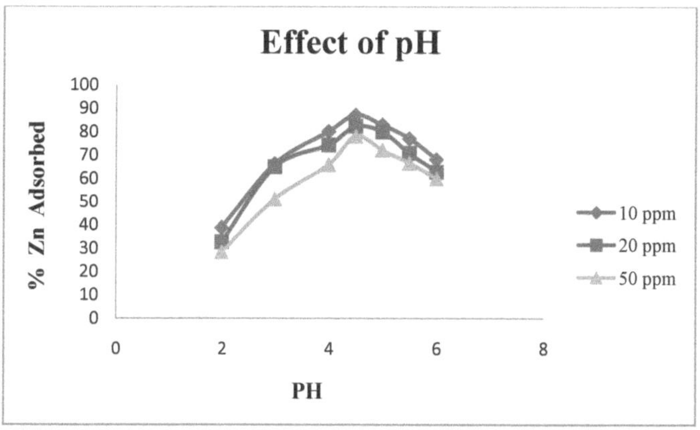

Fig4.21 Effect of pH on biosorption of Zn by *Oscillatoria sp.* (biomass dose=0.1 g, initial Zn ion concentration = 10, 20 and 50 mg/L; temperature = 25 ˚C; agitation speed = 120 rpm; contact time = 120 min).

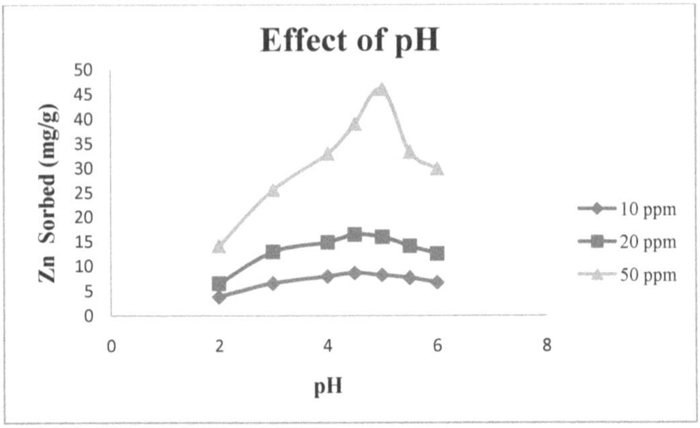

Fig4.22 Effect of pH on biosorption of Zn by *Oscillatoria sp.* (biomass dose=0.1 g, initial Zn ion concentration = 10, 20 and 50 mg/L; temperature = 25 °C; agitation speed = 120 rpm; contact time = 120 min).

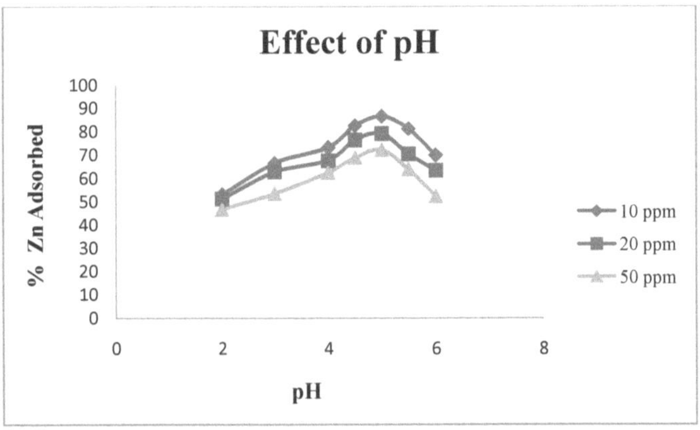

Fig.4.23 Effect of pH on biosorption of Zn by *Spirulina sp.* (biomass dose=0.1 g, initial Zn ion concentration = 10, 20 and 50 mg/L; temperature = 25 °C; agitation speed = 120 rpm; contact time = 120 min).

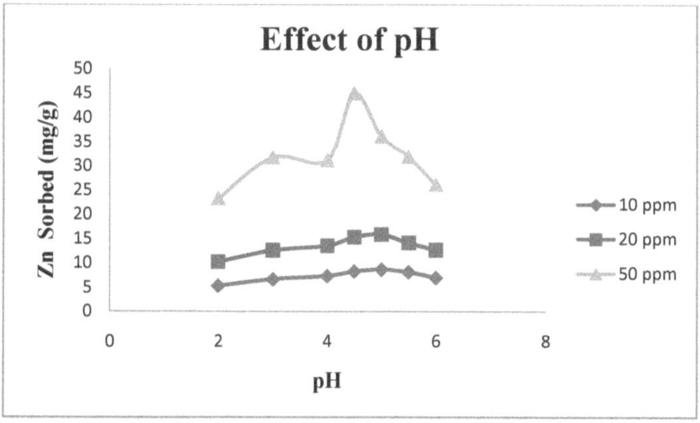

Fig4.24 Effect of pH on biosorption of Zn by *Spirulina sp.* (biomass dose=0.1 g, initial Zn ion concentration = 10, 20 and 50 mg/L; temperature = 25 °C; agitation speed = 120 rpm; contact time = 120 min).

4.5 Effect of contact time for Chromium

Results on the contact time of chromium at different initial metal ion concentrations by *Chlorella pyrenoidosa, Oscillatoria sp.* and *Spirulina sp.* present in the Figures 4.25 to 4.30.

The time required to reach equilibrium for chromium adsorption by *Chlorella pyrenoidosa, Oscillatoria sp.* and *Spirulina sp.* was 60 minute for all initial metal ion concentrations. In the initial concentration of 10 mg/L at the time of 60 minute percentage of remove of Cr was 88.93% , 95.66 %and 82.52% and metal ion uptake capacity was 8.89 mg/g , 9.56 mg/g and 8.25 mg/g and when initial concentration of Cr increased to 100 mg/L , percentage of remove of Cr was 73.25% , 82.67% and 59.57% and uptake capacity was 73.25 mg/g , 82.67 mg/g and 59.57 mg/g for *Chlorella pyrenoidosa, Oscillatoria sp.* and *Spirulina sp.* respectively.

The time taken for Cr adsorption by *Chlorella pyrenoidosa, Oscillatoria sp.* and *Spirulina sp.* was dependent on initial metal ion concentration and increased with increase in concentration of Cr. With increase the initial concentration percentage of biosorption decreased and metal ion uptake capacity was increased.

In effect of time and initial concentration, percentage of remove of Cr and metal ion uptake capacity *for the three biomasses under study has been established as:*

Oscillatoria sp. > *Chlorella pyrenoidosa* > *Spirulina sp.*

When initial concentration increased to 100 mg/L, percentage of remove of Cr and metal ion uptake capacity *for the three biomasses under study has been established as: Oscillatoria sp.* > *Chlorella pyrenoidosa* > *Spirulina sp.*

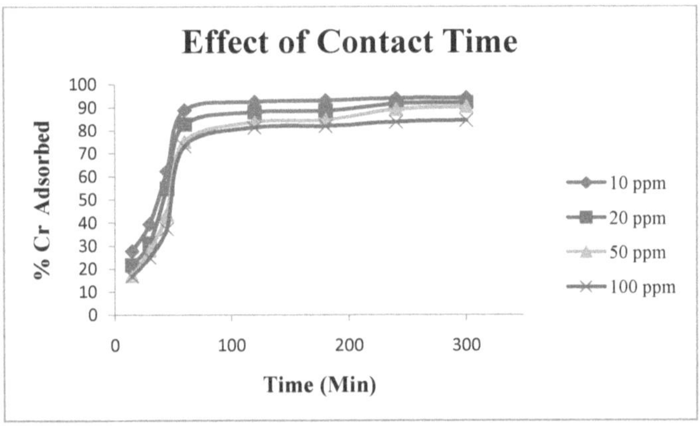

Fig.4.25 Effect of Contact Time on biosorption of Cr by *Chlorella pyrenoidosa* (biomass dose=0.1 g, pH=5, initial Cr ion concentration = 10, 20 50 and 100 mg/L; temperature = 25 ˚C; agitation speed = 120 rpm).

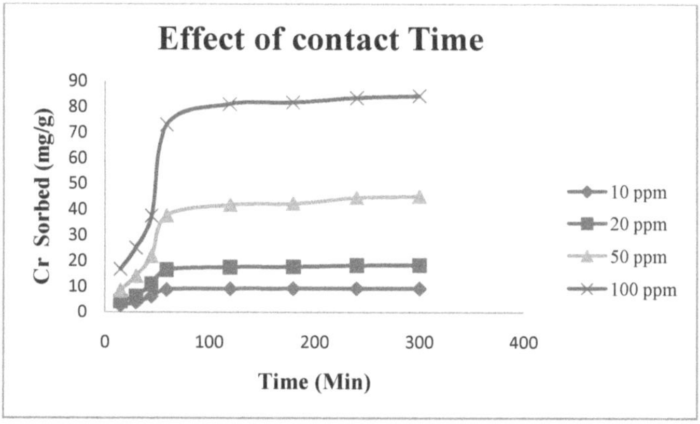

Fig.4.26 Effect of Contact Time on biosorption of Cr by *Chlorella pyrenoidosa* (biomass dose=0.1 g, pH=5, initial Cr ion concentration = 10, 20 50 and 100 mg/L; temperature = 25 ˚C; agitation speed = 120 rpm).

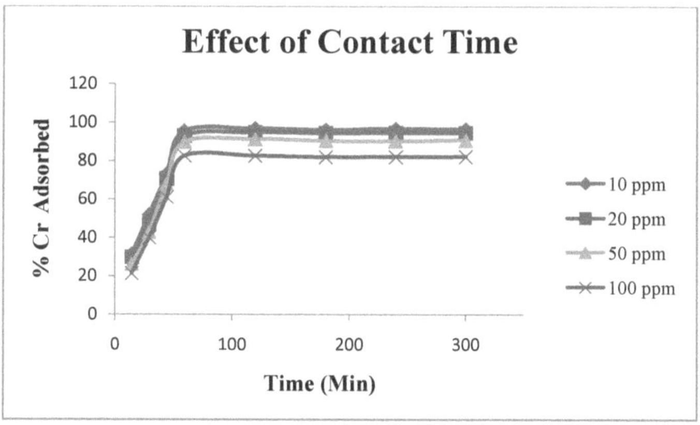

Fig4.27 Effect of Contact Time on biosorption of Cr by *Oscillatoria sp.* (biomass dose=0.1 g, pH=5, initial Cr ion concentration = 10, 20 50 and 100 mg/L; temperature = 25 ˚C; agitation speed = 120 rpm).

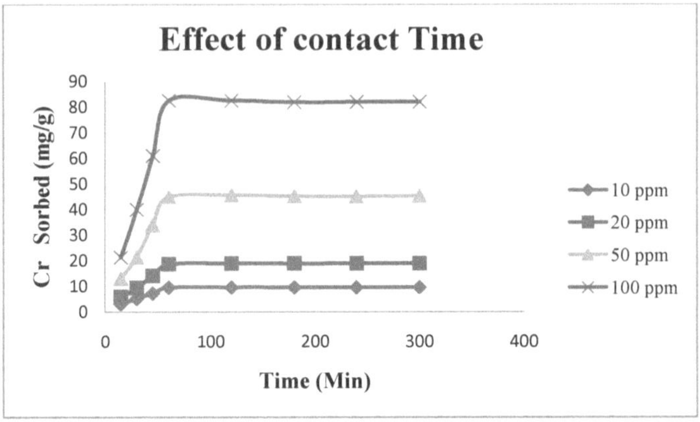

Fig.4.28 Effect of Contact Time on biosorption of Cr by *Oscillatoria sp.* (biomass dose=0.1 g, pH=5, initial Cr ion concentration = 10, 20 50 and 100 mg/L; temperature = 25 °C; agitation speed = 120 rpm).

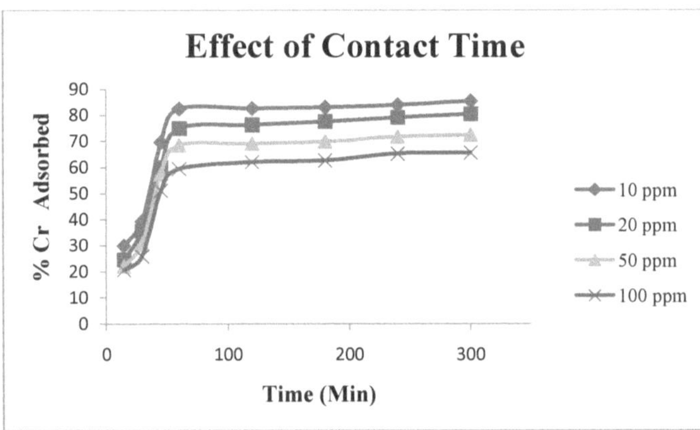

Fig.4.29 Effect of Contact Time on biosorption of Cr by *Spirulina sp.* (biomass dose=0.1 g, pH=5, initial Cr ion concentration = 10, 20 50 and 100 mg/L; temperature = 25 °C; agitation speed = 120 rpm).

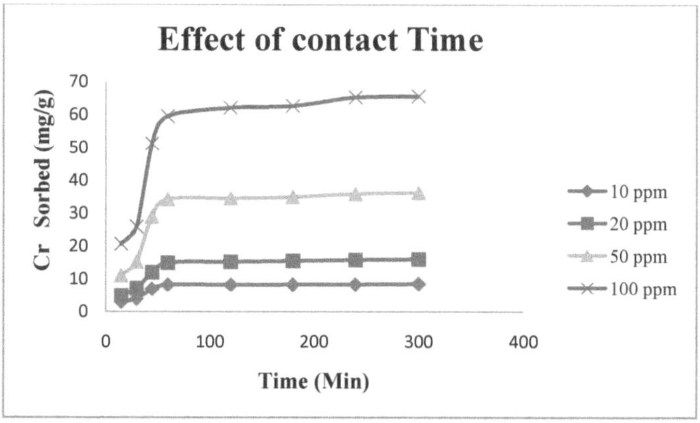

Fig.4.30 Effect of Contact Time on biosorption of Cr by *Spirulina sp.* (biomass dose=0.1 g, pH=5, initial Cr ion concentration = 10, 20 50 and 100 mg/L; temperature = 25 °C; agitation speed = 120 rpm).

4.6 Effect of contact time for Copper

Results on the contact time of copper at different initial metal ion concentrations by *Chlorella pyrenoidosa, Oscillatoria sp.* and *Spirulina sp.* present in the Figures 4.31 to 4.36.

The time of 60 minute was required for all initial metal ion concentration to reach equilibrium for Cu adsorption by *Chlorella pyrenoidosa, Oscillatoria sp.* and *Spirulina sp.* In the initial concentration of 10 mg/L at the time of 60 minute percentage of remove of Cu was 87.36% , 91.80 % and 86.33 % and metal ion uptake capacity was 8.73 mg/g , 9.18 mg/g and 8.63 mg/g and when initial concentration of Cu increased to 100 mg/L , percentage of remove of Cu was 62.60 %,86.30% and 73.67% and uptake capacity was 62.60 mg/g , 86.30 mg/g and 73.67 mg/g for *Chlorella pyrenoidosa, Oscillatoria sp.* and *Spirulina sp.* respectively. The graph (Fig.4.31) shows that time taken for Cu adsorption by *Chlorella pyrenoidosa, Oscillatoria sp.* and *Spirulina sp.* was dependent on initial metal ion concentration and increased with increasing concentration of Cu.

With increase the initial concentration percentage of biosorption decreased and metal ion uptake capacity was increased.

In effect of time and initial concentration, percentage of remove of Cu and metal ion uptake capacity *for the three biomasses under study has been established as:*

Oscillatoria sp. > *Chlorella pyrenoidosa* > *Spirulina sp.*

When initial concentration increased to 100 mg/L, percentage of remove of Cu and metal ion uptake capacity *for the three biomasses under study has been established as:*

Oscillatoria sp. > *Spirulina sp.* > *Chlorella pyrenoidosa*

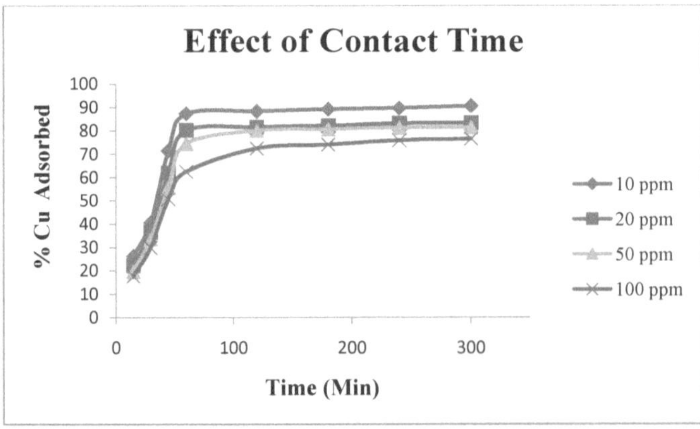

Fig.4.31 Effect of Contact Time on biosorption of Cu by *Chlorella pyrenoidosa* (biomass dose=0.1 g, pH=5, initial Cu ion concentration = 10, 20 50 and 100 mg/L; temperature = 25 °C; agitation speed = 120 rpm).

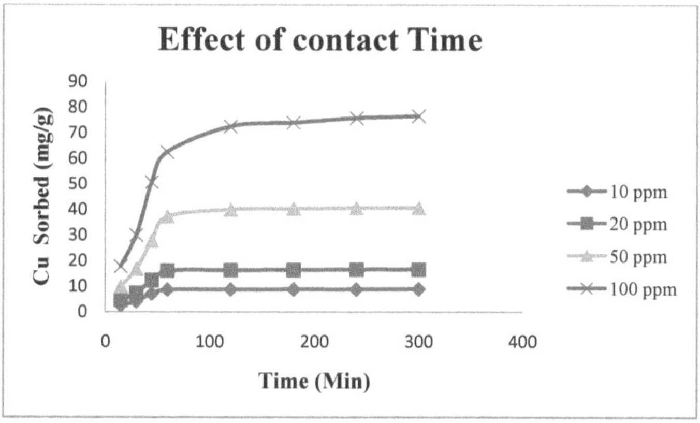

Fig4.32 Effect of Contact Time on biosorption of Cu by *Chlorella pyrenoidosa* (biomass dose=0.1 g, pH=5, initial Cu ion concentration = 10, 20 50 and 100 mg/L; temperature = 25 °C; agitation speed = 120 rpm).

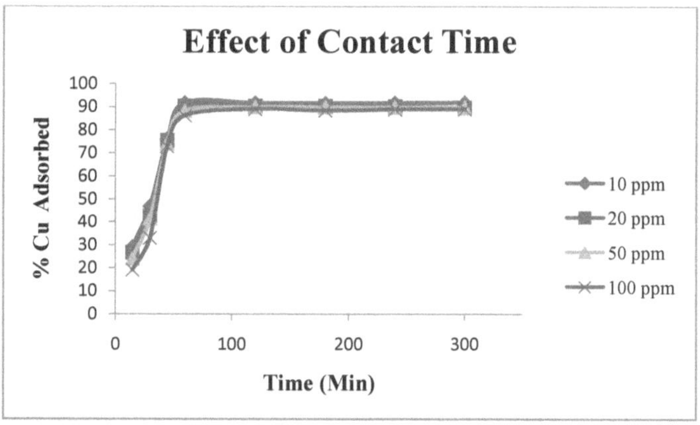

Fig4.33 Effect of Contact Time on biosorption of Cu by *Oscillatoria sp.* (biomass dose=0.1 g, pH=5, initial Cu ion concentration = 10, 20 50 and 100 mg/L; temperature = 25 °C; agitation speed = 120 rpm).

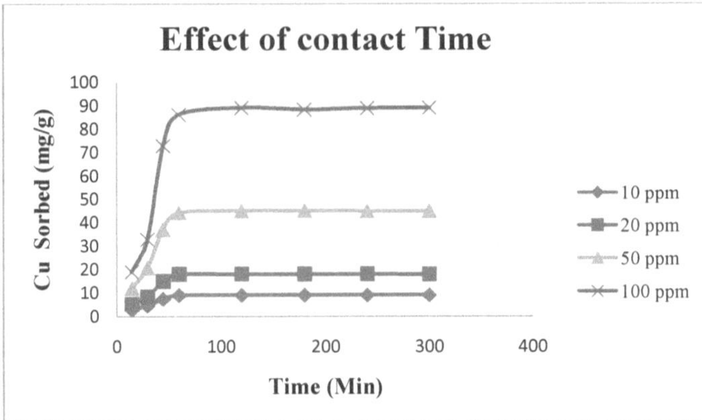

Fig4.34 Effect of Contact Time on biosorption of Cu by *Oscillatoria sp.* (biomass dose=0.1 g, pH=5, initial Cu ion concentration = 10, 20 50 and 100 mg/L; temperature = 25 ˚C; agitation speed = 120 rpm).

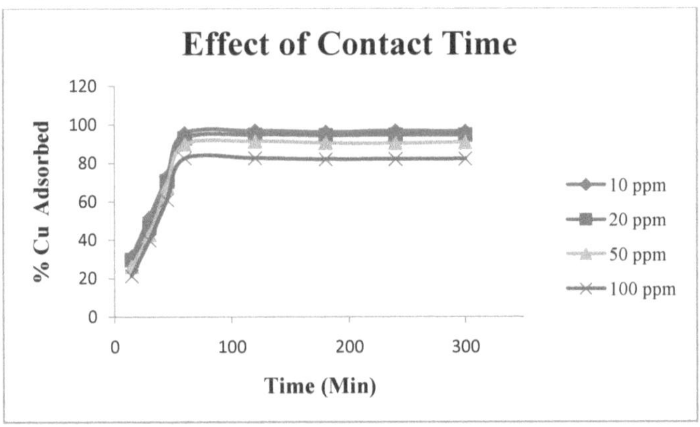

Fig4.35 Effect of Contact Time on biosorption of Cu by *Spirulina sp.* (biomass dose=0.1 g, pH=5, initial Cu ion concentration = 10, 20 50 and 100 mg/L; temperature = 25 ˚C; agitation speed = 120 rpm).

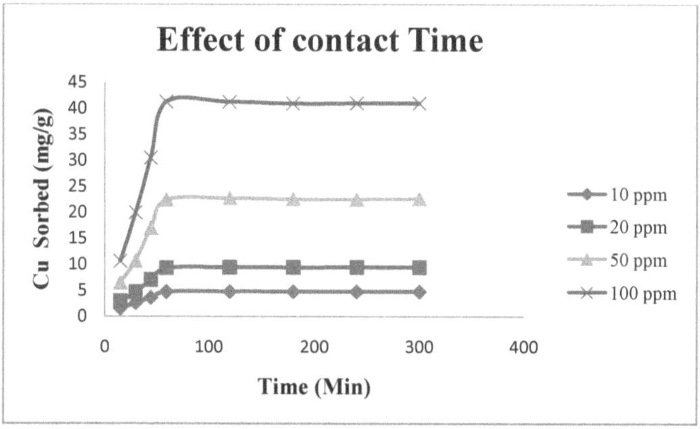

Fig.4.36 Effect of Contact Time on biosorption of Cu by *Spirulina sp.* (biomass dose=0.1 g, pH=5, initial Cu ion concentration = 10, 20 50 and 100 mg/L; temperature = 25 °C; agitation speed = 120 rpm).

4.7 Effect of contact time for Nickel

Results on the contact time of nickel at different initial metal ion concentrations by *Chlorella pyrenoidosa, Oscillatoria sp.* and *Spirulina sp.* present in the Figures 4.37 to 4.42.

The time required to reach equilibrium for Ni adsorption by *Chlorella pyrenoidosa, Oscillatoria sp.* and *Spirulina sp.* was 60 minute for all initial metal ion concentrations. In the initial concentration of 10 mg/L at the time of 60 minute percentage of remove of Ni was 92.07% , 82.33% and 80.25% and metal ion uptake capacity was 9.27 mg/g , 8.23 mg/g and 8.02 mg/g and when initial concentration of Ni increased to 100 mg/L , percentage of remove of Ni was 63.34 %,67.86 % and 63.44 % and uptake capacity was 63.34 mg/g , 67.86 mg/g and 63.44 mg/g for *Chlorella pyrenoidosa, Oscillatoria sp.* and *Spirulina sp.* respectively.

The rate of Ni adsorption by *Chlorella pyrenoidosa*, *Oscillatoria sp.* and *Spirulina sp.* was dependent on initial metal ion concentration . The association is positive as shown by the curve (Fig. 4.38).

As increase the initial concentration percentage of biosorption decreased and metal ion uptake capacity was increased.

In effect of time and initial concentration, percentage of remove of Ni and metal ion uptake capacity *for the three biomasses under study has been established as:*

Chlorella pyrenoidosa > Oscillatoria sp. > Spirulina sp.

And when initial concentration increased to 100 mg/L, percentage of remove of Ni and metal ion uptake capacity *for the three biomasses under study has been established as:*

Oscillatoria sp. > Spirulina sp. > Chlorella pyrenoidosa

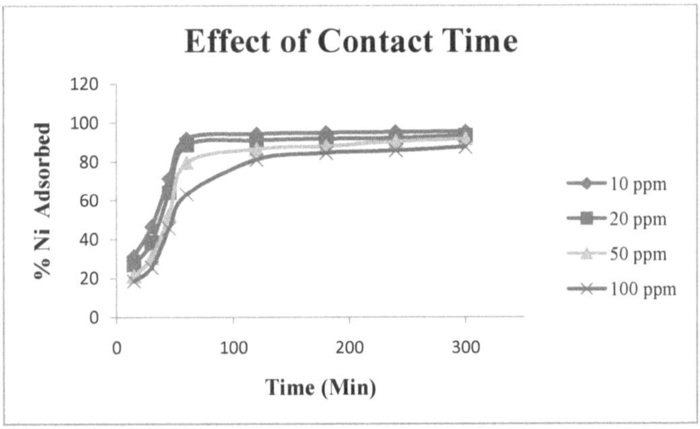

Fig4.37 Effect of Contact Time on biosorption of Ni) by *Chlorella pyrenoidosa* (biomass dose=0.1 g, pH=5, initial Ni ion concentration = 10, 20 50 and 100 mg/L; temperature = 25 °C; agitation speed = 120 rpm).

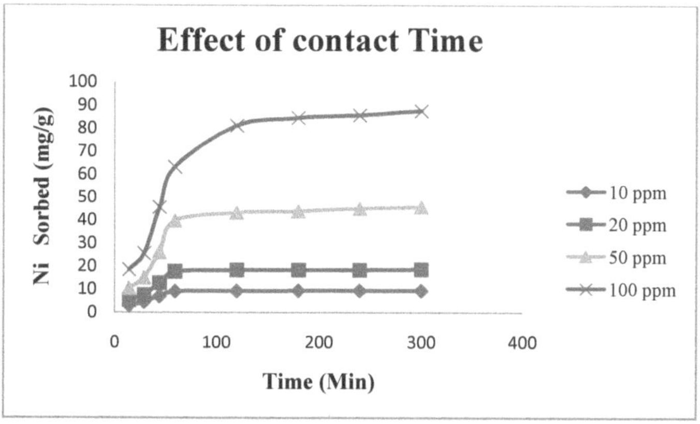

Fig4.38 Effect of Contact Time on biosorption of Ni by *Chlorella pyrenoidosa* (biomass dose=0.1 g, pH=5, initial Ni ion concentration = 10, 20 50 and 100 mg/L; temperature = 25 ˚C; agitation speed = 120 rpm).

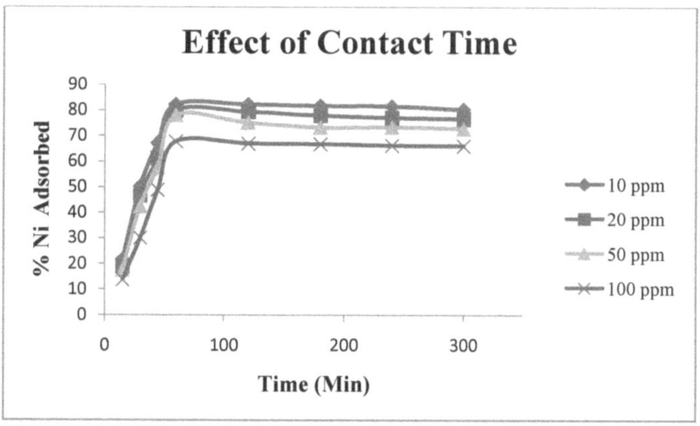

Fig.4.39 Effect of Contact Time on biosorption of Ni by *Oscillatoria sp.* (biomass dose=0.1 g, pH=5, initial Ni ion concentration = 10, 20 50 and 100 mg/L; temperature = 25 ˚C; agitation speed = 120 rpm).

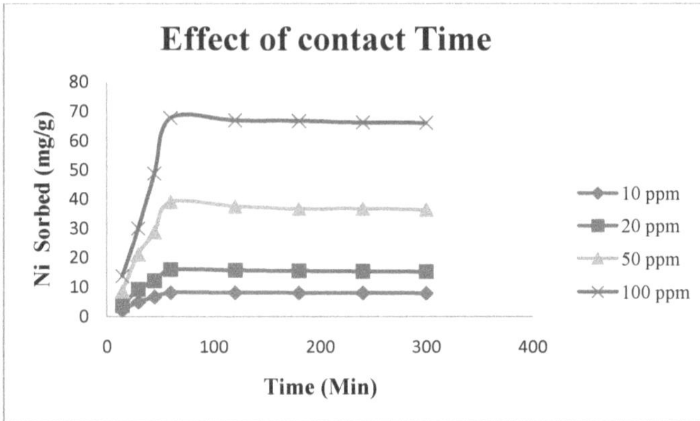

Fig.4.40 Effect of Contact Time on biosorption of Ni by *Oscillatoria sp.* (biomass dose=0.1 g, pH=5, initial Ni ion concentration = 10, 20 50 and 100 mg/L; temperature = 25 °C; agitation speed = 120 rpm).

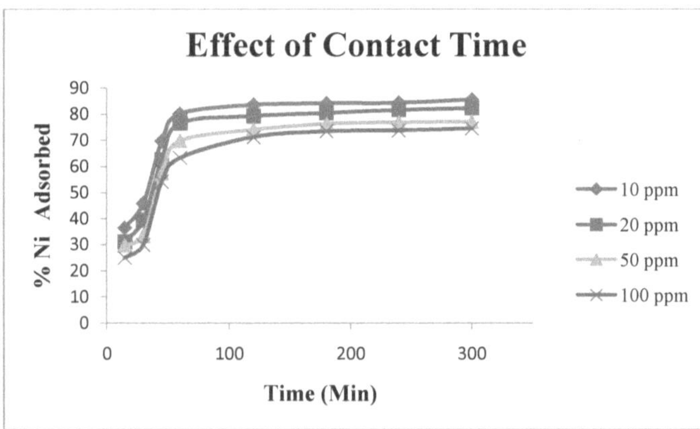

Fig4.41 Effect of Contact Time on biosorption of Ni by *Spirulina sp.* (biomass dose=0.1 g, pH=5, initial Ni ion concentration = 10, 20 50 and 100 mg/L; temperature = 25 °C; agitation speed = 120 rpm).

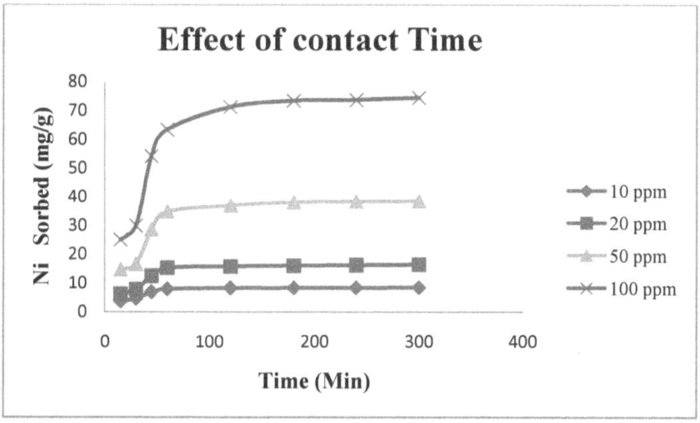

Fig4.42 Effect of Contact Time on biosorption of Ni by *Spirulina sp.* (biomass dose=0.1 g, pH=5, initial Ni ion concentration = 10, 20 50 and 100 mg/L; temperature = 25 °C; agitation speed = 120 rpm).

4.8 Effect of contact time for Zinc

Results on the contact time of Zinc at different initial metal ion concentrations by *Chlorella pyrenoidosa, Oscillatoria sp.* and *Spirulina sp.* present in the Figures 4.43 to 4. 48.

The period required to reach equilibrium for Zn adsorption by *Chlorella pyrenoidosa, Oscillatoria sp.* and *Spirulina sp.* was 60 minute for all initial metal ion concentrations. In the initial concentration of 10 mg/L at the time of 60 minute percentage of remove of Zn was 82.27% , 87.29% and 85.33% and metal ion uptake capacity was 8.22 mg/g , 8.73 mg/g and 8.53 mg/g and when initial concentration of Zn increased to 100 mg/L , percentage of remove of Zn was 64.61% , 72.11% and 67.79% and uptake capacity was 64.61 mg/g , 72.11 mg/g and 67.79 mg/g for *Chlorella pyrenoidosa, Oscillatoria sp.* and *Spirulina sp.* respectively.

The time for Zn adsorption by *Chlorella pyrenoidosa, Oscillatoria sp.* and *Spirulina sp.* was dependent positively on initial metal ion concentration .

With increase the initial concentration percentage of biosorption decreased and metal ion uptake capacity was increased.

In effect of time and initial concentration, percentage of remove of Zn and metal ion uptake capacity *for the three biomasses under study has been established as:*

Oscillatoria sp. > *Spirulina sp.* > *Chlorella pyrenoidosa*

And when initial concentration increased to 100 mg/L, percentage of remove of Zn and metal ion uptake capacity *for the three biomasses under study has been established as:*

Oscillatoria sp. > *Spirulina sp.* > *Chlorella pyrenoidosa*

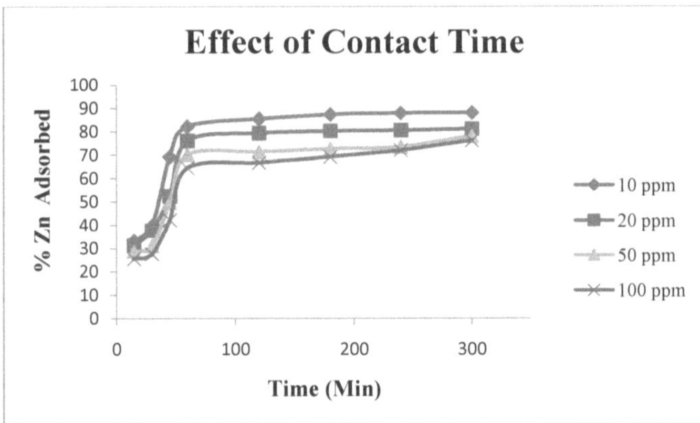

Fig.4.43 Effect of Contact Time on biosorption of Zn by *Chlorella pyrenoidosa* (biomass dose=0.1 g, pH=5, initial Zn ion concentration = 10, 20 50 and 100 mg/L; temperature = 25 °C; agitation speed = 120 rpm).

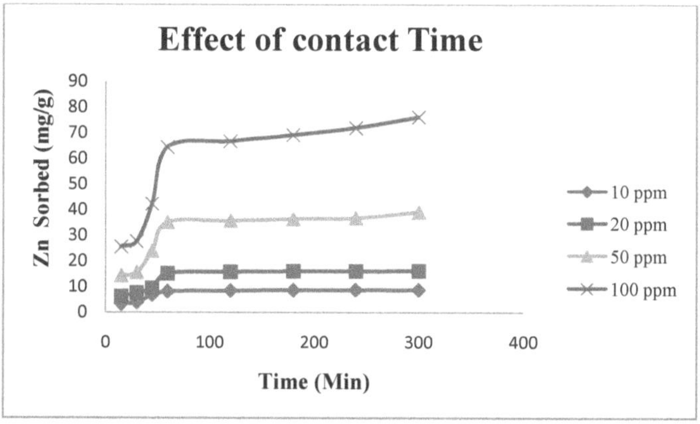

Fig.4.44 Effect of Contact Time on biosorption of Zn by *Chlorella pyrenoidosa* (biomass dose=0.1 g, pH=5, initial Zn ion concentration = 10, 20 50 and 100 mg/L; temperature = 25 ˚C; agitation speed

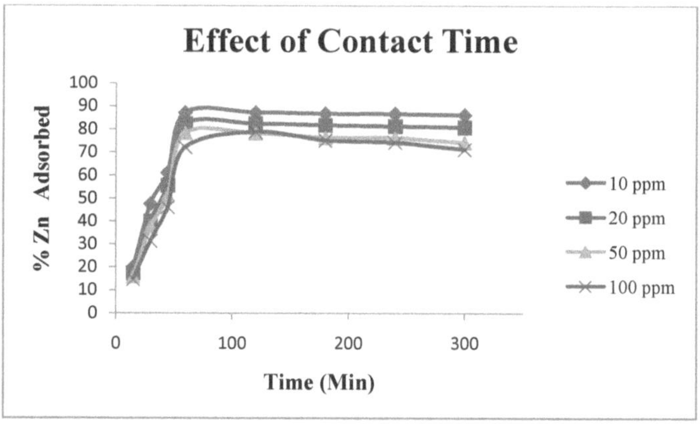

Fig4.45 Effect of Contact Time on biosorption of Zn by *Oscillatoria sp.* (biomass dose=0.1 g, pH=5, initial Zn ion concentration = 10, 20 50 and 100 mg/L; temperature = 25 ˚C; agitation speed = 120 rpm).

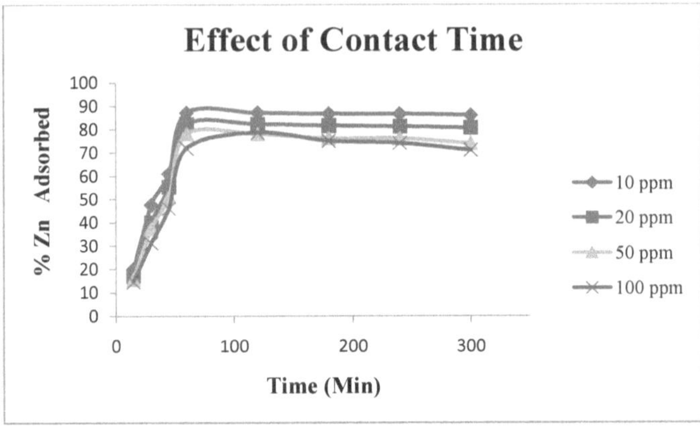

Fig.4.46 Effect of Contact Time on biosorption of Zn by *Oscillatoria sp.* (biomass dose=0.1 g, pH=5, initial Zn ion concentration = 10, 20 50 and 100 mg/L; temperature = 25 ˚C; agitation speed = 120 rpm).

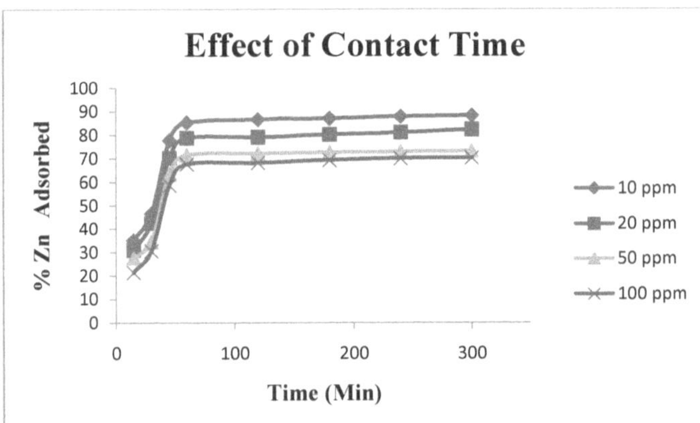

Fig4.47 Effect of Contact Time on biosorption of Zn by *Spirulina sp.* (biomass dose=0.1 g, pH=5, initial Zn ion concentration = 10, 20 50 and 100 mg/L; temperature = 25 ˚C; agitation speed = 120 rpm).

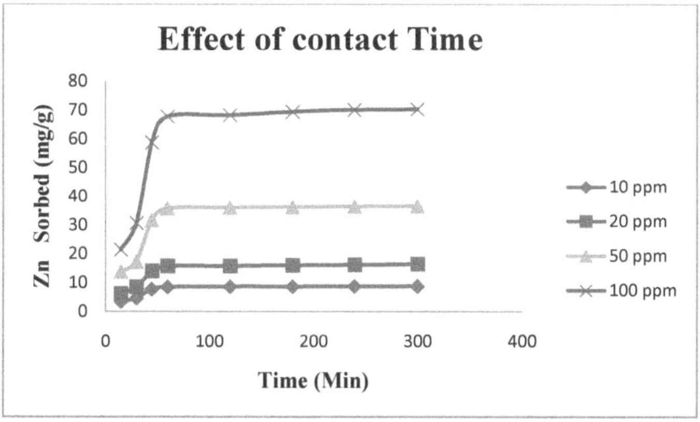

Fig4.48 Effect of Contact Time on biosorption of Zn by *Spirulina sp.* (biomass dose=0.1 g, pH=5, initial Zn ion concentration = 10, 20 50 and 100 mg/L; temperature = 25 ˚C; agitation speed = 120 rpm).

4.9 Effect of adsorbent dosage for Chromium

Results on the effect of biomass dosage of Cr at different initial metal ion concentrations by *Chlorella pyrenoidosa, Oscillatoria sp.* and *Spirulina sp.* present in the 4.49 to 4.54.

Percentage of biosorption accursed at the initial concentration of 10 mg/L at the time of 120 minute at pH= 5 and 0.1± 0.001 g of biomass for Cr was 92.51% , 96.81% and 82.67 % and metal ion uptake capacity was 9.25 mg/g , 4.84 mg/g and 8.26 mg/g and when initial concentration of Cr increased to 50 mg/L , percentage of biosortion of Cr was 83.64% , 91.21% and 69.12% and uptake capacity was 41.82 mg/g , 22.80 mg/g and 34.56 mg/g for *Chlorella pyrenoidosa, Oscillatoria sp.* and *Spirulina sp.* respectively .

When amount of biomass increased from 0.1 ± 0.001 g to 3 ± 0.001 g , percentage of biosorption accursed at the initial concentration of 10 mg/L at the time of 120 minute at pH= 5 for Cr was 99.82% , 99.37% and 92.57% and metal ion uptake capacity was 0.332 mg/g , 0.330 mg/g and 0.308 mg/g and when initial concentration of Cr increased to 50 mg/L , percentage of biosortion of Cr was 95.66% , 96.87% and 86.35% and uptake capacity was 1.59 mg/g , 1.61mg/g and 1.44 mg/g for *Chlorella pyrenoidosa, Oscillatoria sp.* and *Spirulina sp.* respectively.

With increase the initial concentration percentage of biosorption decreased and metal ion uptake capacity was increased. With increased the amount of biomass observed that percentage of biosorption increased and metal ion uptake capacity was decreased.

In effect of biomass dosage and initial concentration of 10 mg/L, percentage of remove of Cr and metal ion uptake capacity *for the three biomass under study has been established as: Chlorella pyrenoidosa > Oscillatoria sp. > Spirulina sp.*

And when initial concentration increased to 50 mg/L, percentage of remove of Cr and metal ion uptake capacity *for the three biomasses under study has been established as:*

Oscillatoria sp. > Chlorella pyrenoidosa > Spirulina sp.

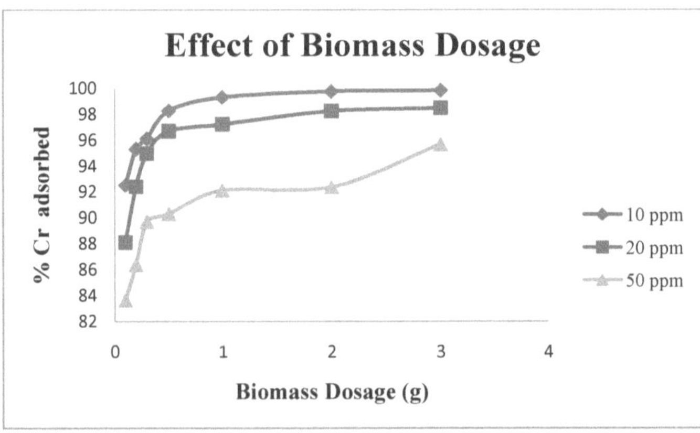

Fig4.49 Effect of Biomass Dosage on biosorption of Cr by *Chlorella pyrenoidosa* (pH=5, initial Cr ion concentration = 10, 20 and 50 mg/L; temperature = 25 °C; agitation speed = 120 rpm; contact time = 120 min).

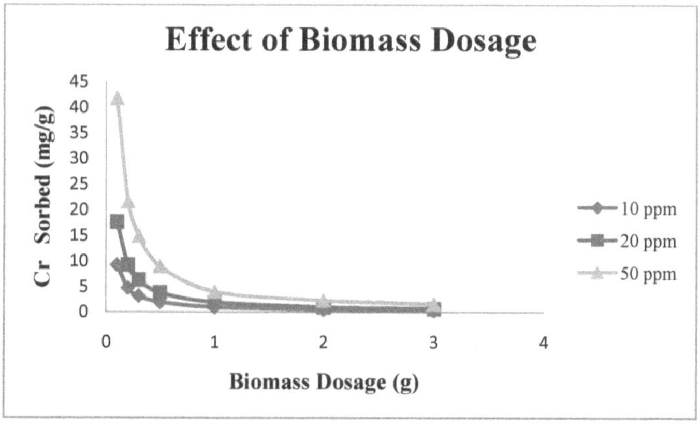

Fig4.50 Effect of Biomass Dosage on biosorption of Cr by *Chlorella pyrenoidosa* (pH=5, initial Cr ion concentration = 10, 20 and 50 mg/L; temperature = 25 ˚C; agitation speed = 120 rpm; contact time = 120 min).

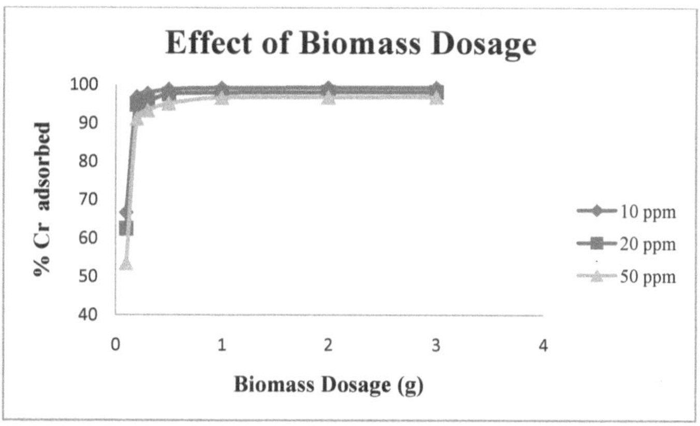

Fig4.51 Effect of Biomass Dosage on biosorption of Cr by *Oscillatoria sp.* (pH=5, initial Cr ion concentration = 10, 20 and 50 mg/L; temperature = 25 ˚C; agitation speed = 120 rpm; contact time = 120 min).

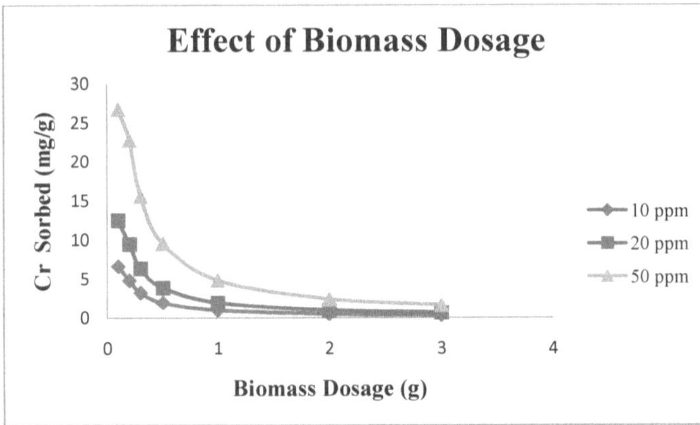

Fig4.52 Effect of Biomass Dosage on biosorption of Cr by *Oscillatoria sp.* (pH=5, initial Cr ion concentration = 10, 20 and 50 mg/L; temperature = 25 °C; agitation speed = 120 rpm; contact time = 120 min).

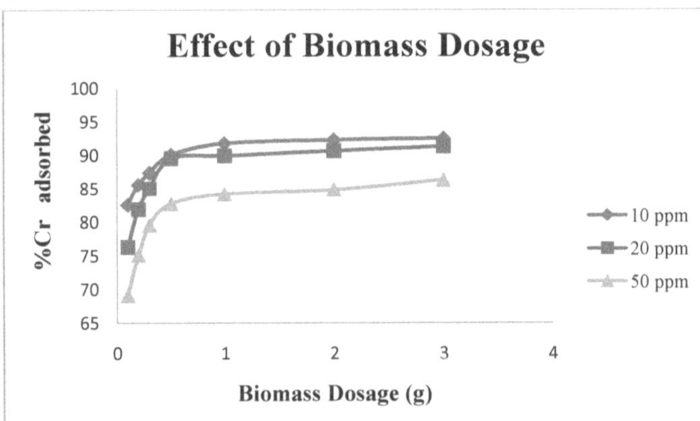

Fig4.53 Effect of Biomass Dosage on biosorption of Cr by *Spirulina sp.* (pH=5, initial Cr ion concentration = 10, 20 and 50 mg/L; temperature = 25 °C; agitation speed = 120 rpm; contact time = 120 min).

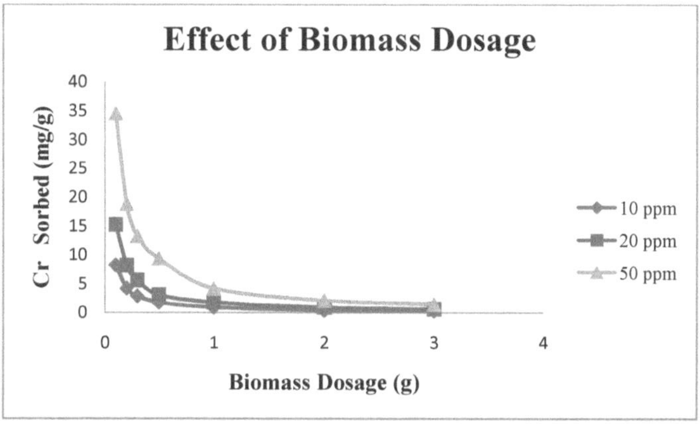

Fig4.54 Effect of Biomass Dosage on biosorption of Cr by *Spirulina sp.* (pH=5, initial Cr ion concentration = 10, 20 and 50 mg/L; temperature = 25 °C; agitation speed = 120 rpm; contact time = 120 min).

4.10 Effect of adsorbent dosage for Copper

Results on the effect of biomass dosage of Cu at different initial metal ion concentrations by *Chlorella pyrenoidosa, Oscillatoria sp.* and *Spirulina sp.* present in the Figures 4.55 to 4.60. percentage of biosorption accursed at the initial concentration of 10 mg/L at the time of 120 minute at pH= 5 and 0.1± 0.001 g of biomass for Cu was 88.48% , 91.70 % and 90.09% and metal ion uptake capacity was 8.85 mg/g , 9.17 mg/g and 9.01 mg/g and when initial concentration of Cu increased to 50 mg/L , percentage of biosortion of Cu was 80.13% , 89.95% and 84.44% and uptake capacity was 40.06 mg/g , 44.97 mg/g and 42.22 mg/g for *Chlorella pyrenoidosa, Oscillatoria sp.* and *Spirulina sp.* respectively. When amount of biomass increased from 0.1 ± 0.001 g to 3 ± 0.001 g , percentage of biosorption accursed at the initial concentration of 10 mg/L at the time of 120 minute at pH= 5 for Cu was 98.76% , 99.20% and 99.35% and metal ion uptake capacity was 0.329 mg/g , 0.330 mg/g and 0.331 mg/g and when initial concentration of Cu increased to 50 mg/L , percentage of biosortion of Cu was 94.66% , 95.66% and 95.78% and uptake capacity was 1.57 mg/g , 1.59 mg/g and 1.60 mg/g for *Chlorella pyrenoidosa, Oscillatoria sp.* and *Spirulina sp.* respectively.

With increase the initial concentration percentage of biosorption decreased and metal ion uptake capacity was increased. With increased the amount of biomass observed that percentage of biosorption increased and metal ion uptake capacity was decreased.

In effect of biomass dosage and initial concentration of 10 mg/L, percentage of remove of Cu and metal ion uptake capacity *for the three biomasses under study has been established as:* *Spirulina sp.* > *Oscillatoria sp.* > *Chlorella pyrenoidosa*

And when initial concentration increased to 50 mg/L, percentage of remove of Cu and metal ion uptake capacity *for the three biomasses under study has been established as:*

Spirulina sp. > *Oscillatoria sp.* > *Chlorella pyrenoidosa*

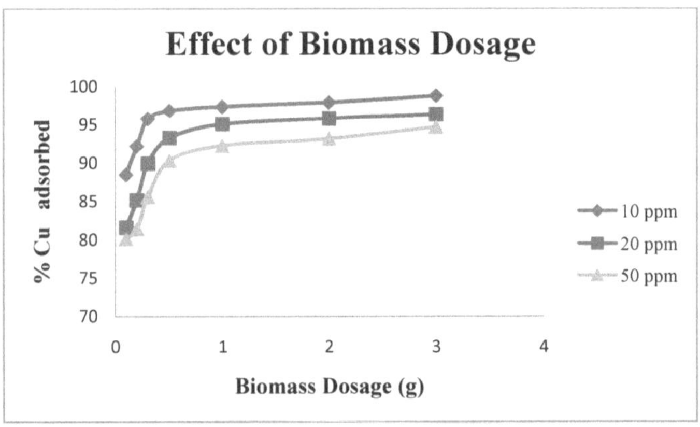

Fig4.55 Effect of Biomass Dosage on biosorption of Cu by *Chlorella pyrenoidosa* (pH=5, initial Cu ion concentration = 10, 20 and 50 mg/L; temperature = 25 ˚C; agitation speed = 120 rpm; contact time = 120 min).

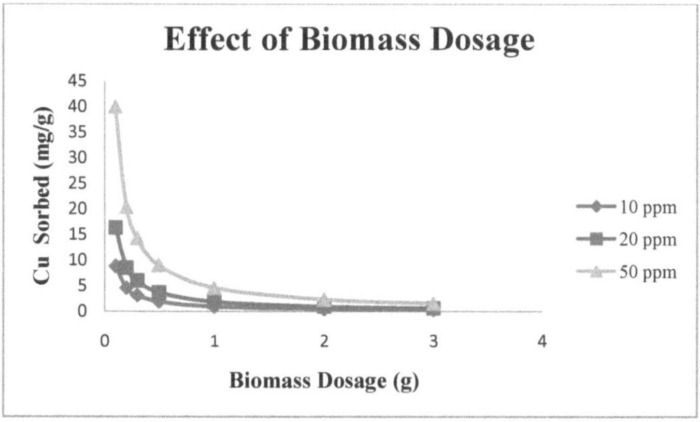

Fig4.56 Effect of Biomass Dosage on biosorption of Cu by *Chlorella pyrenoidosa* (pH=5, initial Cu ion concentration = 10, 20 and 50 mg/L; temperature = 25 °C; agitation speed = 120 rpm; contact time = 120 min).

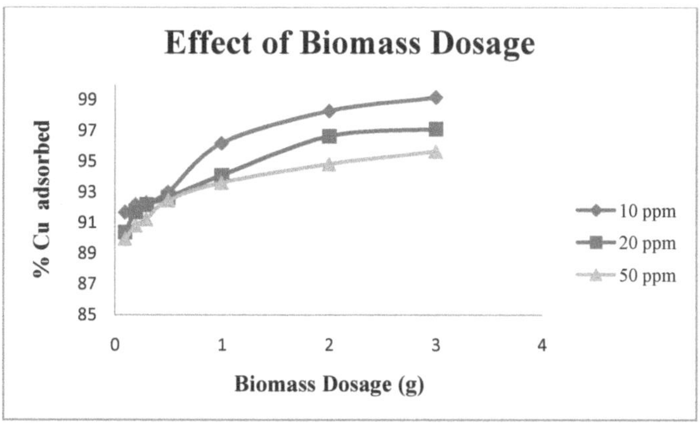

Fig4.57 Effect of Biomass Dosage on biosorption of Cu by *Oscillatoria sp.* (pH=5, initial Cu ion concentration = 10, 20 and 50 mg/L; temperature = 25 °C; agitation speed = 120 rpm; contact time = 120 min).

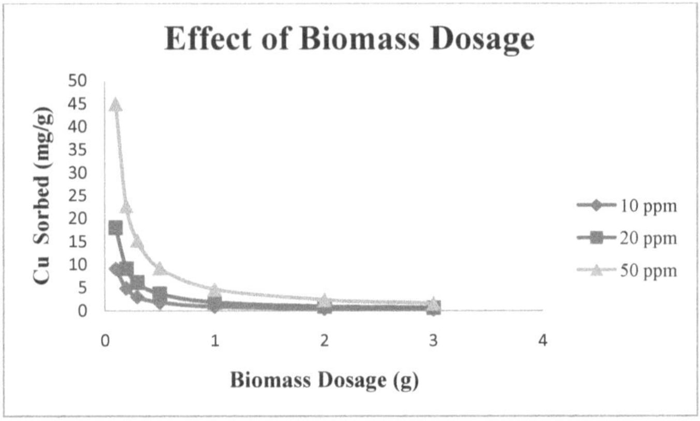

Fig4.58 Effect of Biomass Dosage on biosorption of Cu by *Oscillatoria sp.* (pH=5, initial Cu ion concentration = 10, 20 and 50 mg/L; temperature = 25 °C; agitation speed = 120 rpm; contact time = 120 min).

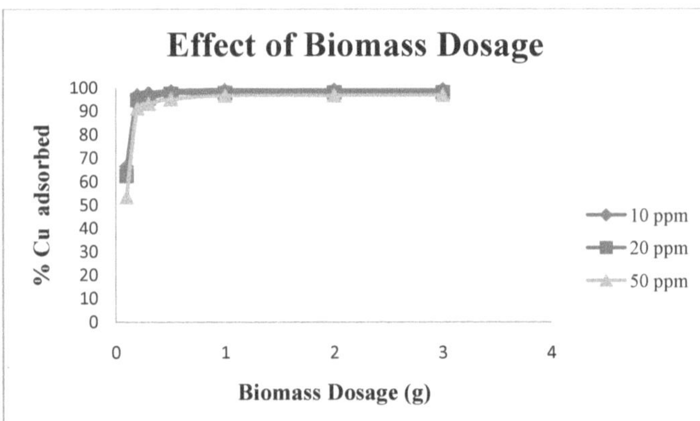

Fig4.59 Effect of Biomass Dosage on biosorption of Cu by *Spirulina sp.* (pH=5, initial Cu ion concentration = 10, 20 and 50 mg/L; temperature = 25 °C; agitation speed = 120 rpm; contact time = 120 min).

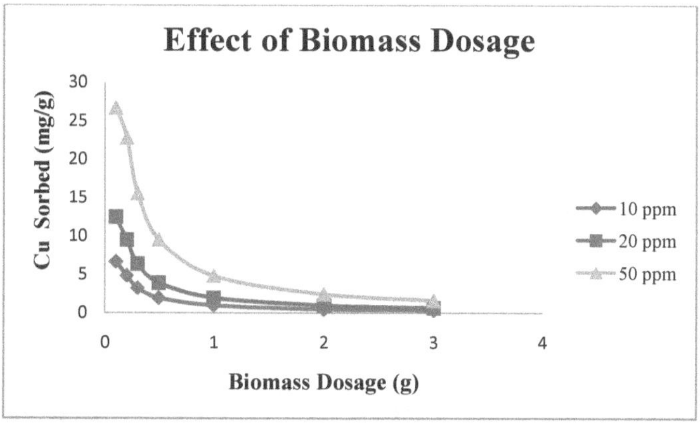

Fig4.60 Effect of Biomass Dosage on biosorption of Cu by *Spirulina sp.* (pH=5, initial Cu ion concentration = 10, 20 and 50 mg/L; temperature = 25 ˚C; agitation speed = 120 rpm; contact time = 120 min).

4.11 Effect of adsorbent dosage for Nickel

Results on the effect of biomass dosage of Ni at different initial metal ion concentrations by *Chlorella pyrenoidosa, Oscillatoria sp.* and *Spirulina sp.* present in the Figures4.61 to4.66.

percentage of biosorption accursed at the initial concentration of 10 mg/L at the time of 120 minute at pH= 5 and 0.1± 0.001 g of biomass for Ni was 94.31% , 82.17% and 83.60% and metal ion uptake capacity was 9.43 mg/g , 8.22 mg/g and 8.36 mg/g and when initial concentration of Ni increased to 50 mg/L , percentage of biosortion of Ni was 86.60% , 75.12% and 74.05% and uptake capacity was 43.30 mg/g , 37.56 mg/g and 37.02 mg/g for *Chlorella pyrenoidosa, Oscillatoria sp.* and *Spirulina sp.* respectively . When amount of biomass increased from 0.1 ± 0.001 g to 3 ± 0.001 g , percentage of biosorption accursed at the initial concentration of 10 mg/L at the time of 120 minute at pH= 5 for Ni was 93.10% , 99.37% and 90.64 % and metal ion uptake capacity was 0.332 mg/g , 0.310 mg/g and 0.302 mg/g and when initial concentration of Ni increased to 50 mg/L , percentage of biosortion of Ni was 91.92% , 90.19 % and

85.01% and uptake capacity was 1.53 mg/g , 1.50 mg/g and 1.44 mg/g for *Chlorella pyrenoidosa, Oscillatoria sp.* and *Spirulina sp.* respectively.

With increase the initial concentration percentage of biosorption decreased and metal ion uptake capacity was increased. With increased the amount of biomass observed that percentage of biosorption increased and metal ion uptake capacity was decreased.
In effect of biomass dosage and initial concentration of 10 mg/L, percentage of remove of Ni and metal ion uptake capacity *for the three biomasses under study has been established as:* *Oscillatoria sp.* > *Chlorella pyrenoidosa* > *Spirulina sp.*

And when initial concentration increased to 50 mg/L, percentage of remove of Ni and metal ion uptake capacity *for the three biomasses under study has been established as:*

Chlorella pyrenoidosa > *Oscillatoria sp.* > *Spirulina sp.*

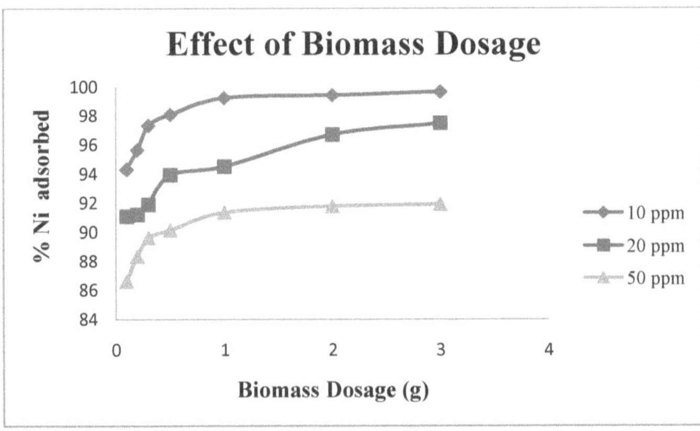

Fig4.61 Effect of Biomass Dosage on biosorption of Ni by *Chlorella pyrenoidosa* (pH=5, initial Ni ion concentration = 10, 20 and 50 mg/L; temperature = 25 ˚C; agitation speed = 120 rpm; contact time = 120 min).

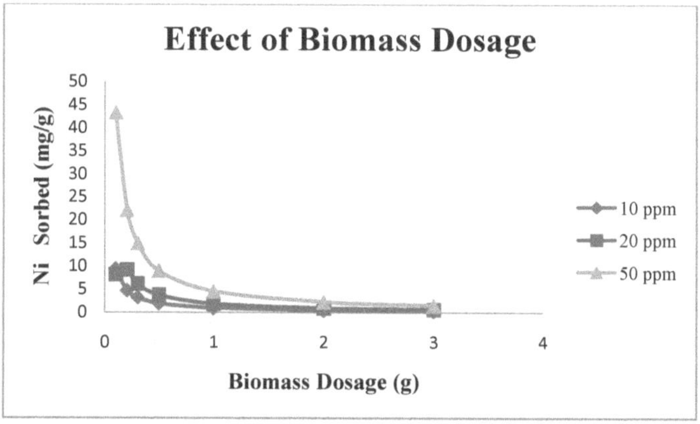

Fig4.62 Effect of Biomass Dosage on biosorption of Ni by *Chlorella pyrenoidosa* (pH=5, initial Ni ion concentration = 10, 20 and 50 mg/L; temperature = 25 ˚C; agitation speed = 120 rpm; contact time = 120 min).

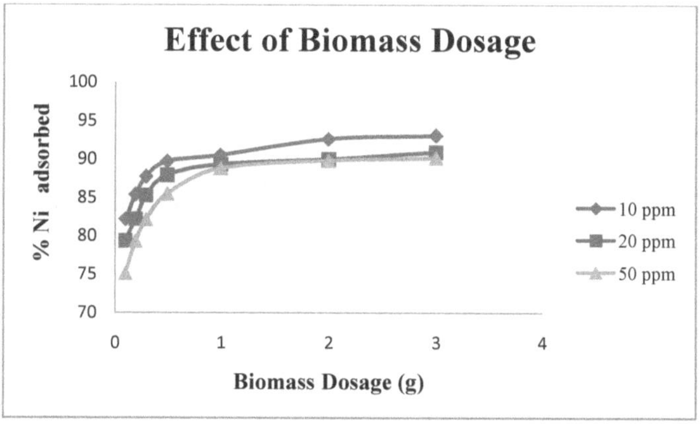

Fig4.63 Effect of Biomass Dosage on biosorption of Ni by *Oscillatoria sp.* (pH=5, initial Ni ion concentration = 10, 20 and 50 mg/L; temperature = 25 ˚C; agitation speed = 120 rpm; contact time = 120 min).

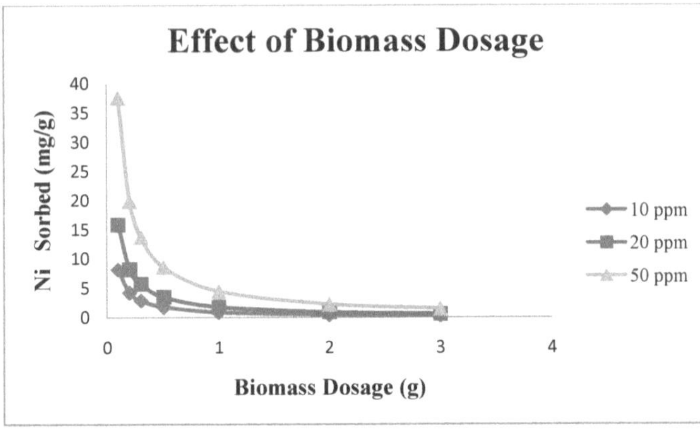

Fig4.64 Effect of Biomass Dosage on biosorption of Ni by *Oscillatoria sp.* (pH=5, initial Ni ion concentration = 10, 20 and 50 mg/L; temperature = 25 ˚C; agitation speed = 120 rpm; contact time = 120 min).

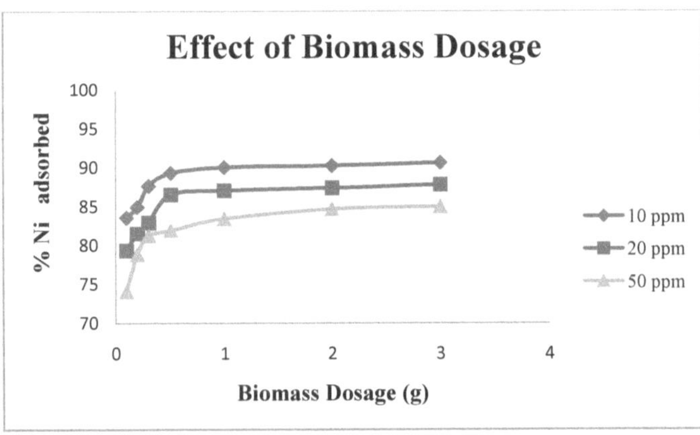

Fig4.65 Effect of Biomass Dosage on biosorption of Ni by *Spirulina sp.* (pH=5, initial Ni ion concentration = 10, 20 and 50 mg/L; temperature = 25 ˚C; agitation speed = 120 rpm; contact time = 120 min).

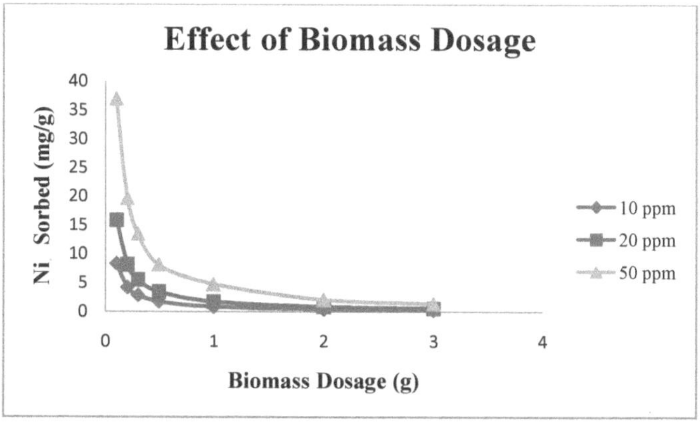

Fig4.66 Effect of Biomass Dosage on biosorption of Ni by *Spirulina sp.* (pH=5, initial Ni ion concentration = 10, 20 and 50 mg/L; temperature = 25 ˚C; agitation speed = 120 rpm; contact time = 120 min).

4.12 Effect of adsorbent dosage for Zinc

Results on the effect of biomass dosage of Zn at different initial metal ion concentrations by *Chlorella pyrenoidosa, Oscillatoria sp.* and *Spirulina sp.* present in the Figures 4.67 to 4.72.

percentage of biosorption accursed at the initial concentration of 10 mg/L at the time of 120 minute at pH= 5 and 0.1± 0.001 g of biomass for Zn was 85.55% , 87.29% and 86.69% and metal ion uptake capacity was 8.56 mg/g , 8.82 mg/g and 8.67 mg/g and when initial concentration of Zn increased to 50 mg/L , percentage of biosortion of Zn was 71.33% , 78.13% and 72.23% and uptake capacity was 35.66 mg/g , 39.07 mg/g and 36.11 mg/g for *Chlorella pyrenoidosa, Oscillatoria sp.* and *Spirulina sp.* respectively. When amount of biomass increased from 0.1 ± 0.001 g to 3 ± 0.001 g , percentage of biosorption accursed at the initial concentration of 10 mg/L at the time of 120 minute at pH= 5 for Zn was 97.45% , 97.33% and 98.93% and metal ion uptake capacity was 0.325 mg/g , 0.320 mg/g and 0.329 mg/g and when initial concentration of Zn increased to 50 mg/L , percentage of biosortion of Zn was 91.55% , 93.35% and

92.80% and uptake capacity was 1.52mg/g , 1.55 mg/g and 1.54 mg/g for *Chlorella pyrenoidosa, Oscillatoria sp.* and *Spirulina sp.* respectively.

With increase the initial concentration percentage of biosorption decreased and metal ion uptake capacity was increased. With increased the amount of biomass observed that percentage of biosorption increased and metal ion uptake capacity was decreased.

In effect of biomass dosage and initial concentration of 10 mg/L, percentage of remove of Zn and metal ion uptake capacity *for the three biomasses under study has been established as: Spirulina sp. > Chlorella pyrenoidosa > Oscillatoria sp.*

And when initial concentration increased to 50 mg/L, percentage of remove of Zn and metal ion uptake capacity *for the three biomasses under study has been established as: Oscillatoria sp. > Spirulina sp. > Chlorella pyrenoidosa*

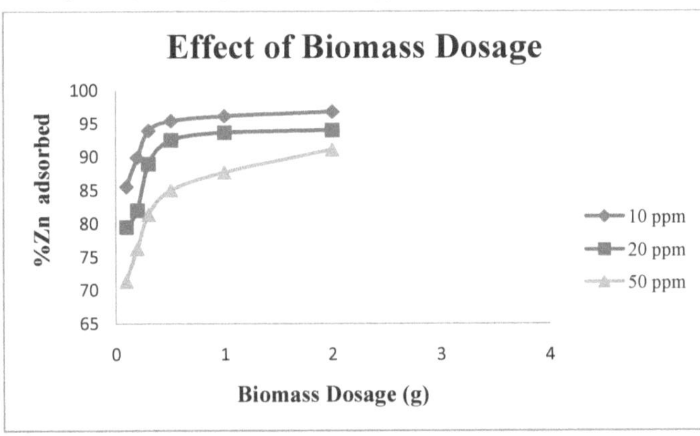

Fig4.67 Effect of Biomass Dosage on biosorption of Zn by *Chlorella pyrenoidosa* (pH=5, initial Zn ion concentration = 10, 20 and 50 mg/L; temperature = 25 °C; agitation speed = 120 rpm; contact time = 120 min).

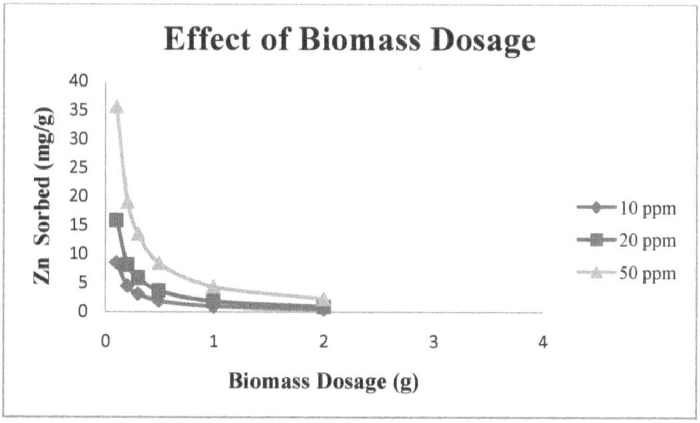

Fig4.68 Effect of Biomass Dosage on biosorption of Zn by *Chlorella pyrenoidosa* (pH=5, initial Zn ion concentration = 10, 20 and 50 mg/L; temperature = 25 ˚C; agitation speed = 120 rpm; contact time = 120 min).

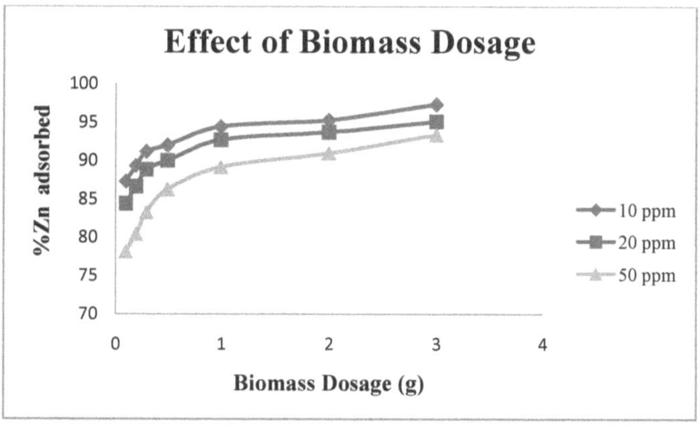

Fig4.69 Effect of Biomass Dosage on biosorption of Zn by *Oscillatoria sp.* (pH=5, initial Zn ion concentration = 10, 20 and 50 mg/L; temperature = 25 ˚C; agitation speed = 120 rpm; contact time = 120 min).

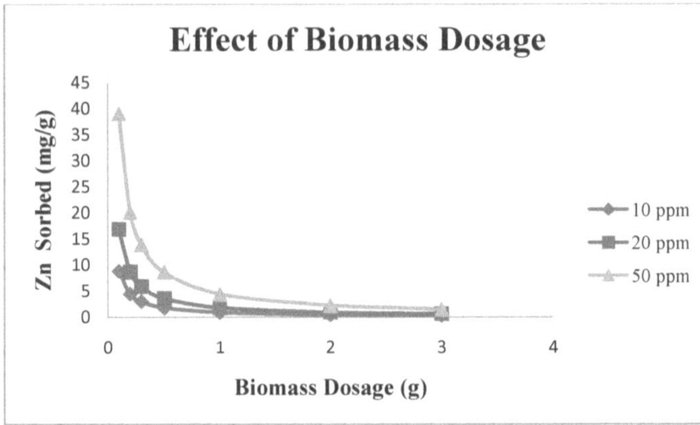

Fig4.70 Effect of Biomass Dosage on biosorption of Zn by *Oscillatoria sp.* (pH=5, initial Zn ion concentration = 10, 20 and 50 mg/L; temperature = 25 ˚C; agitation speed = 120 rpm; contact time = 120 min).

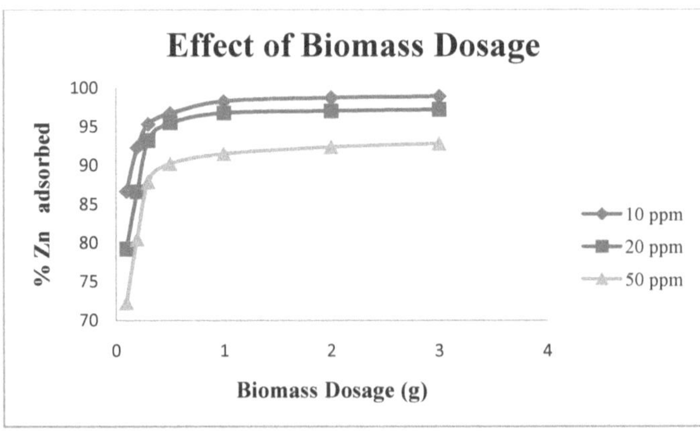

Fig4.71 Effect of Biomass Dosage on biosorption of Zn by *Spirulina sp.* (pH=5, initial Zn ion concentration = 10, 20 and 50 mg/L; temperature = 25 ˚C; agitation speed = 120 rpm; contact time = 120 min).

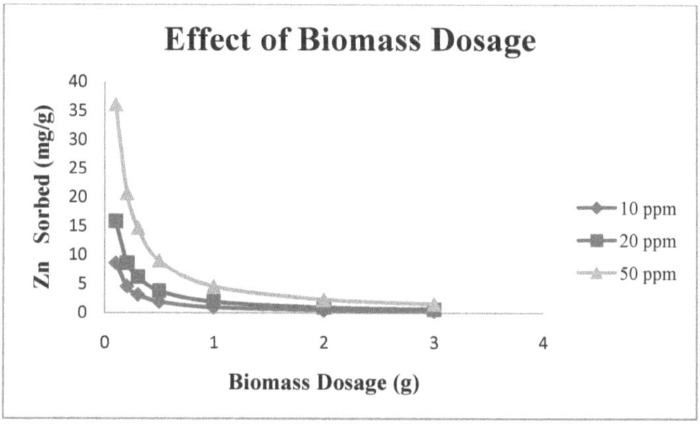

Fig4.72 Effect of Biomass Dosage on biosorption of Zn by *Spirulina sp.* (pH=5, initial Zn ion concentration = 10, 20 and 50 mg/L; temperature = 25 ˚C; agitation speed = 120 rpm; contact time = 120 min).

4.13 Effect of temperature for Chromium

Results on the effect of temperature of Cr at initial metal ion concentration of 50 mg/L by *Chlorella pyrenoidosa, Oscillatoria sp.* and *Spirulina sp.* present in the Figures 4.73 to 4.78.

Maximum percentage of biosorption accursed at the initial concentration of 50 mg/L at the time of 120 minute at pH= 5 for Cr was 82.36%, 91.62% and 72.83% and metal ion uptake capacity was 41.18 mg/g , 22.90 mg/g and 36.41 mg/g at the temperature of 30 °C ,35 °C and 40 °C for *Chlorella pyrenoidosa, Oscillatoria sp.* and *Spirulina sp.* respectively.

The findings of *Chlorella pyrenoidosa* indicate that the sorption percentage increased with increase in temperature up to 30 ° C and there was a decrease in sorption percentage with further increase in temperature. This may be due to the shrinkage of cells at higher temperature.

The findings of *Oscillatoria sp.* indicate that the temperature has no significant effect on the biosorption of Cr over tested 10-40 °C at pH= 5. Biosorption to be a passive energy independent process.

The findings of *Spirulina sp.* indicate that the sorption percentage increased with increase in temperature up to 40 ° C.

In effect of temperature and initial concentration of 50 mg/L, percentage of remove of Cr and metal ion uptake capacity *for the three biomass under study has been established as:*

Oscillatoria sp. > *Chlorella pyrenoidosa* > *Spirulina sp.*

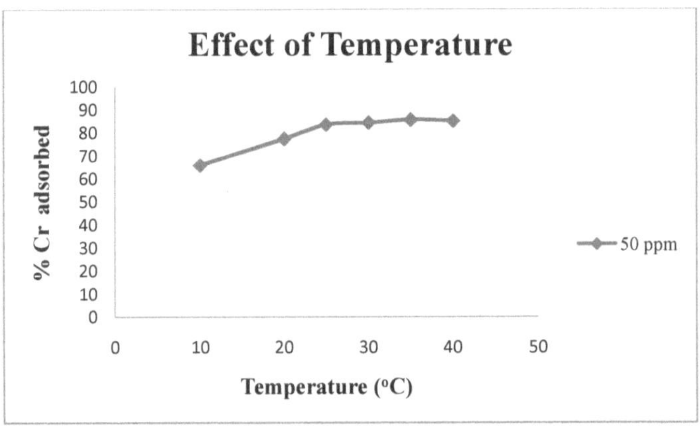

Fig.4.73 Effect of Temperature on biosorption of Cr by *Chlorella pyrenoidosa* (biomass dose=0.1 g, pH=5, initial Cr ion concentration = 50 mg/L; temperature = 25 °C; agitation speed = 120 rpm; contact time = 120 min).

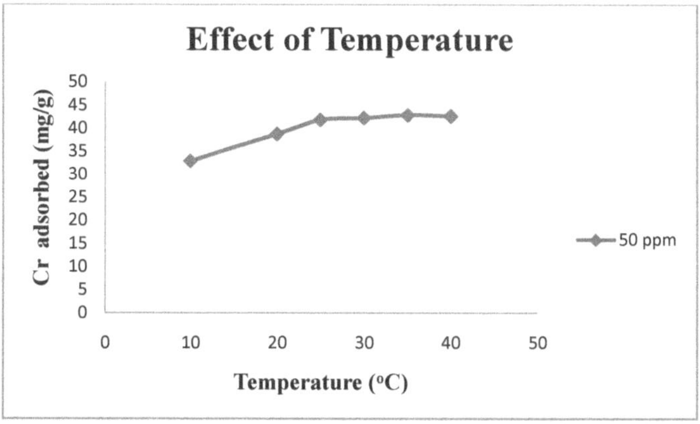

Fig4.74 Effect of Temperature on biosorption of Cr by *Chlorella pyrenoidosa* (biomass dose=0.1 g, pH=5, initial Cr ion concentration = 50 mg/L; temperature = 25 ◦C; agitation speed = 120 rpm; contact time = 120 min).

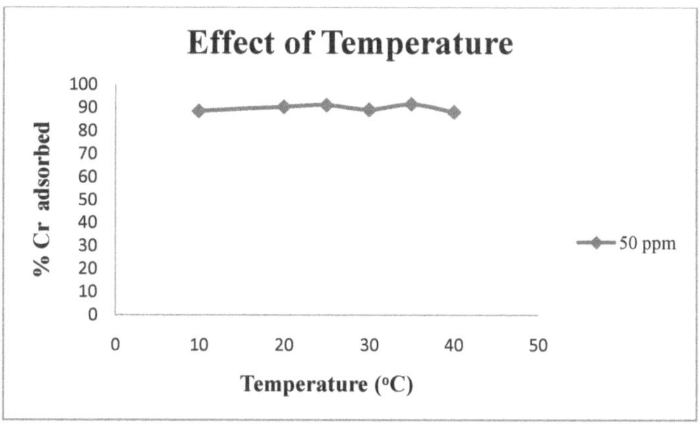

Fig4.75 Effect of Temperature on biosorption of Cr by *Oscillatoria sp.* (biomass dose=0.1 g, pH=5, initial Cr ion concentration = 50 mg/L; temperature = 25 ◦C; agitation speed = 120 rpm; contact time = 120 min).

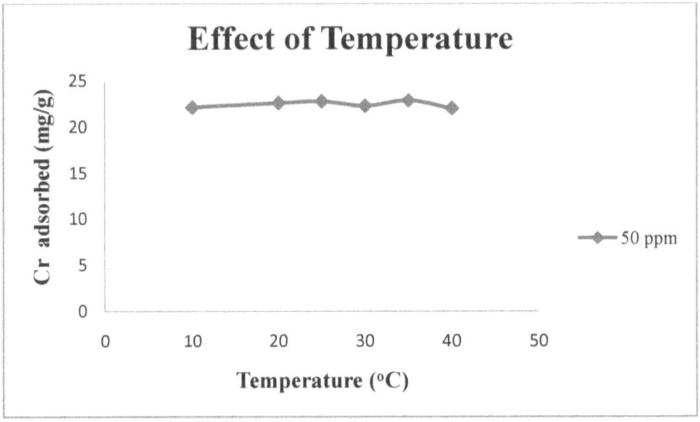

Fig4.76 Effect of Temperature on biosorption of Cr by *Oscillatoria sp.* (biomass dose=0.1 g, pH=5, initial Cr ion concentration = 50 mg/L; temperature = 25 ◦C; agitation speed = 120 rpm; contact time = 120 min).

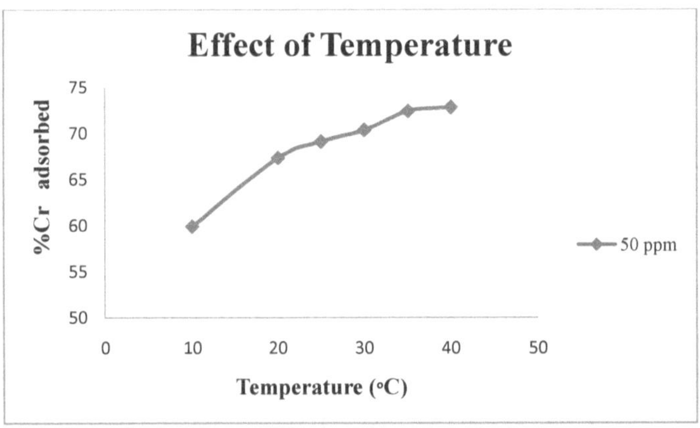

Fig4.77 Effect of Temperature on biosorption of Cr by *Spirulina sp.* (biomass dose=0.1 g, pH=5, initial Cr ion concentration = 50 mg/L; temperature = 25 ◦C; agitation speed = 120 rpm; contact time = 120 min).

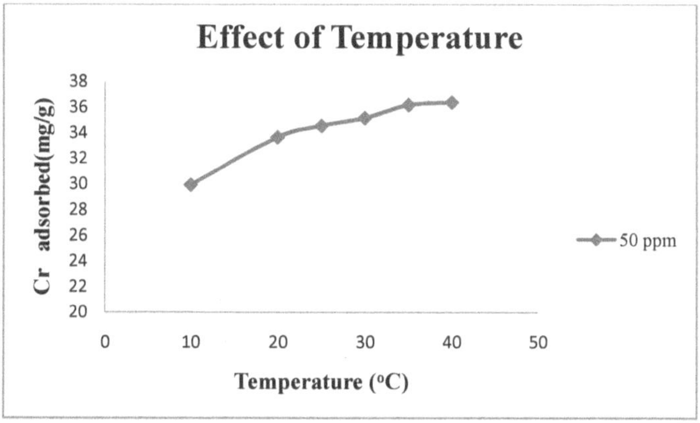

Fig4.78 Effect of Temperature on biosorption of Cr by *Spirulina sp.* (biomass dose=0.1 g, pH=5, initial Cr ion concentration = 50 mg/L; temperature = 25 ◦C; agitation speed = 120 rpm; contact time = 120 min).

4.14 Effect of temperature for Copper

Results on the effect of temperature of Cu at initial metal ion concentration of 50 mg/L by *Chlorella pyrenoidosa, Oscillatoria sp.* and *Spirulina sp.* present in the Figures 4.79 to 4.84.

Maximum percentage of biosorption accursed at the initial concentration of 50 mg/L at the time of 120 minute at pH= 5 for Cu was 82.84%, 87.20% and 89.36% and metal ion uptake capacity was 41.42 mg/g, 43.60 mg/g and 36.41 mg/g at the temperature of 40 °C, 35 °C and 40 °C for *Chlorella pyrenoidosa, Oscillatoria sp.* and *Spirulina sp.* respectively.

The findings of *Chlorella pyrenoidosa* indicate that the sorption percentage increased with increase in temperature up to 40 ° C.

The findings of *Oscillatoria sp.* indicate that the temperature has no significant effect on the biosorption of Cu over tested 10-40 °C at pH= 5. Biosorption to be a passive energy independent process.

The findings of *Spirulina sp.* indicate that the sorption percentage increased with increase in temperature up to 40 ° C but increase the sorption percentage from 25 – 40 °C was very low.

In effect of temperature and initial concentration of 50 mg/L, percentage of remove of Cu and metal ion uptake capacity *for the three biomasses under study has been established as:*
Spirulina sp. > *Oscillatoria sp.* > *Chlorella pyrenoidosa*

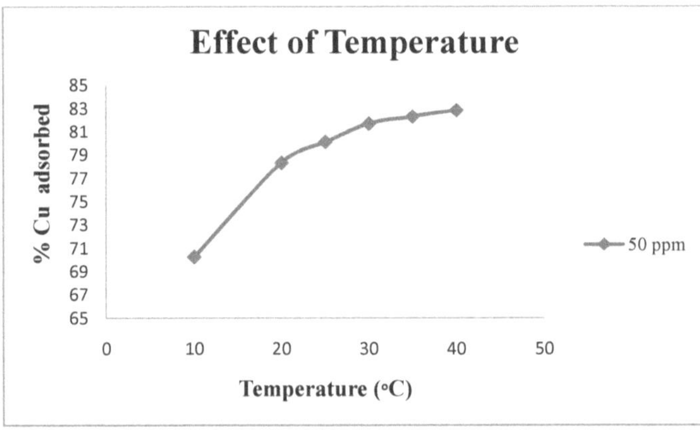

Fig4.79 Effect of Temperature on biosorption of Cu by *Chlorella pyrenoidosa* (biomass dose=0.1 g, pH=5, initial Cu ion concentration = 50 mg/L; temperature = 25 °C; agitation speed = 120 rpm; contact time = 120 min).

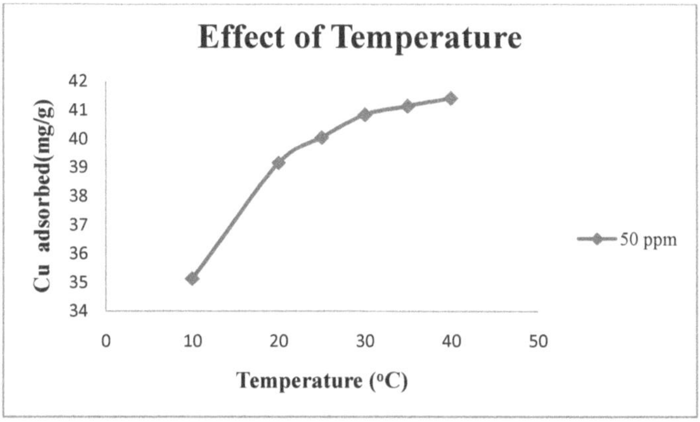

Fig4.80 Effect of Temperature on biosorption of Cu by *Chlorella pyrenoidosa* (biomass dose=0.1 g, pH=5, initial Cu ion concentration = 50 mg/L; temperature = 25 ◦C; agitation speed = 120 rpm; contact time = 120 min).

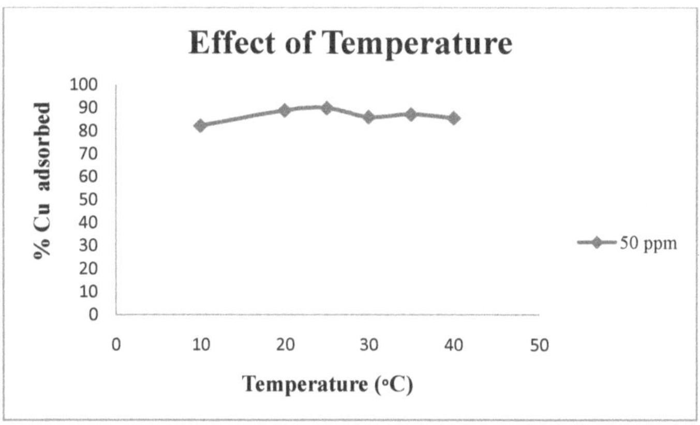

Fig4.81 Effect of Temperature on biosorption of Cu by *Oscillatoria sp.* (biomass dose=0.1 g, pH=5, initial Cu ion concentration = 50 mg/L; temperature = 25 ◦C; agitation speed = 120 rpm; contact time = 120 min).

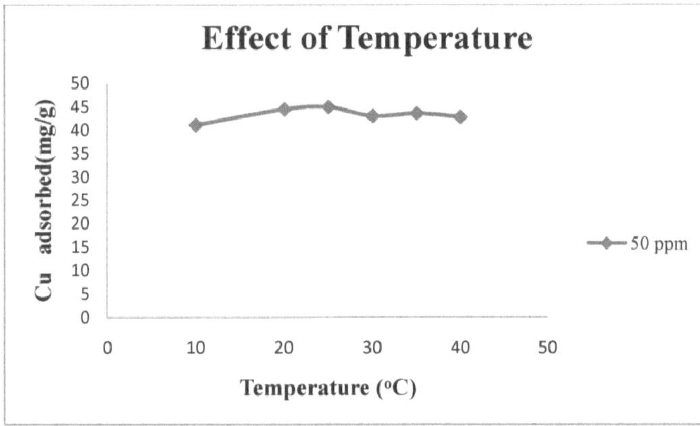

Fig4.82 Effect of Temperature on biosorption of Cu by *Oscillatoria sp.* (biomass dose=0.1 g, pH=5, initial Cu ion concentration = 50 mg/L; temperature = 25 ◦C; agitation speed = 120 rpm; contact time = 120 min).

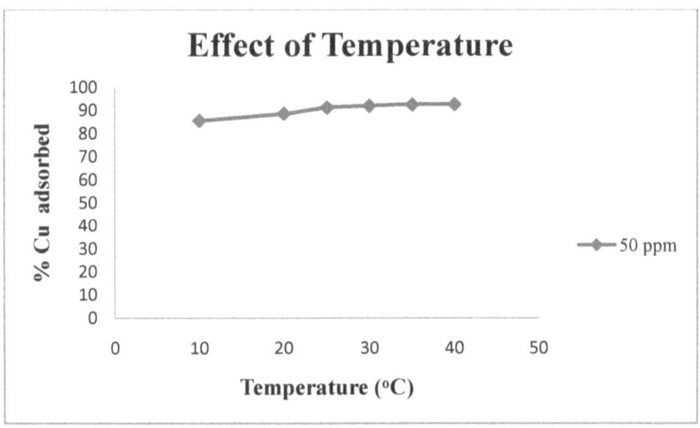

Fig4.83 Effect of Temperature on biosorption of Cu by *Spirulina sp.* (biomass dose=0.1 g, pH=5, initial Cu ion concentration = 50 mg/L; temperature = 25 ◦C; agitation speed = 120 rpm; contact time = 120 min).

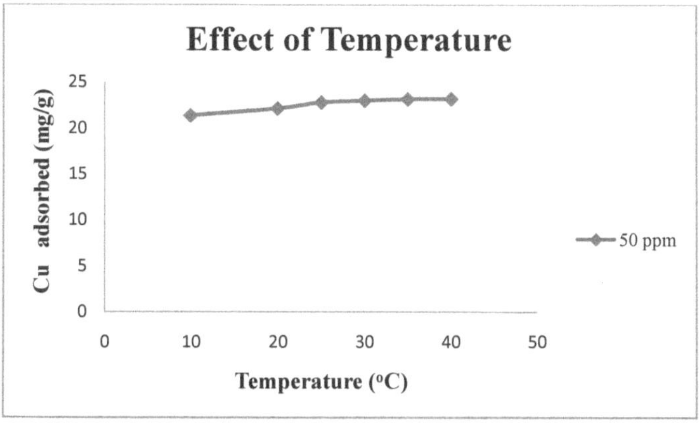

Fig4.84 Effect of Temperature on biosorption of Cu by *Spirulina sp.* (biomass dose=0.1 g, pH=5, initial Cu ion concentration = 50 mg/L; temperature = 25 °C; agitation speed = 120 rpm; contact time = 120 min).

4.15 Effect of temperature for Nickel

Results on the effect of temperature of Ni at initial metal ion concentration of 50 mg/L by *Chlorella pyrenoidosa, Oscillatoria sp.* and *Spirulina sp.* present in the Figures 4.85 to 4.90.

Maximum percentage of biosorption accursed at the initial concentration of 50 mg/L at the time of 120 minute at pH= 5 for Ni was 89.42%, 76.29% and 76.78% and metal ion uptake capacity was 44.71 mg/g, 38.14 mg/g and 38.39 mg/g at the temperature of 40 °C ,30 °C and 40 °C for *Chlorella pyrenoidosa, Oscillatoria sp.* and *Spirulina sp.* respectively.

The findings of *Chlorella pyrenoidosa* indicate that the sorption percentage increased with increase in temperature up to 40 ° C.

The findings of *Oscillatoria sp.* indicate that the temperature has no significant effect on the biosorption of Ni over tested 10-40 °C at pH= 5. Biosorption to be a passive energy independent process.

The findings of *Spirulina sp.* indicate that the sorption percentage increased with increase in temperature up to 40 ° C.

In effect of temperature and initial concentration of 50 mg/L, percentage of remove of Ni and metal ion uptake capacity *for the three biomass under study has been established as:*

Chlorella pyrenoidosa > *Spirulina sp.* > *Oscillatoria sp.*

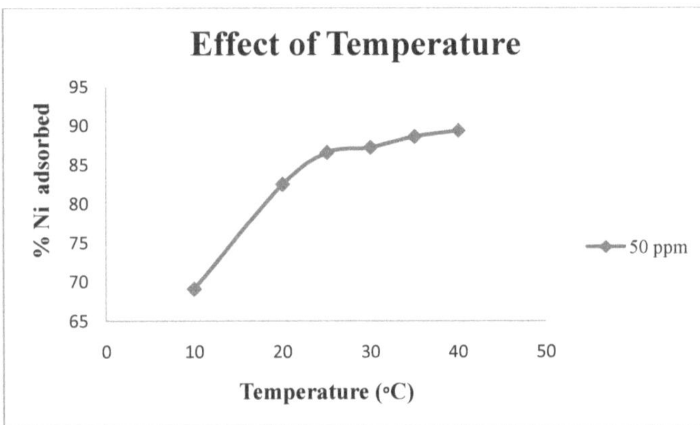

Fig4.85 Effect of Temperature on biosorption of Ni by *Chlorella pyrenoidosa* (biomass dose=0.1 g, pH=5, initial Ni ion concentration = 50 mg/L; temperature = 25 ◦C; agitation speed = 120 rpm; contact time = 120 min).

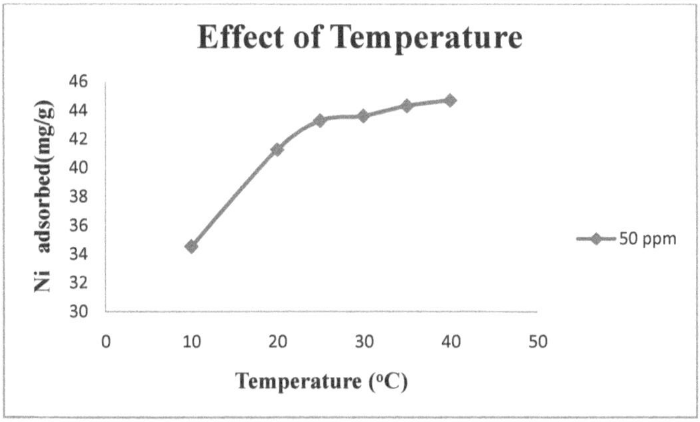

Fig4.86 Effect of Temperature on biosorption of Ni by *Chlorella pyrenoidosa* (biomass dose=0.1 g, pH=5, initial Ni ion concentration = 50 mg/L; temperature = 25 ∘C; agitation speed = 120 rpm; contact time = 120 min).

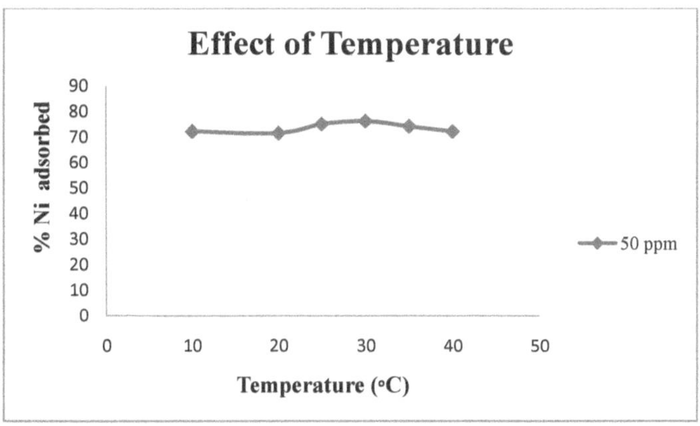

Fig4.87 Effect of Temperature on biosorption of Ni by *Oscillatoria sp.* (biomass dose=0.1 g, pH=5, initial Cu ion concentration = 50 mg/L; temperature = 25 ∘C; agitation speed = 120 rpm; contact time = 120 min).

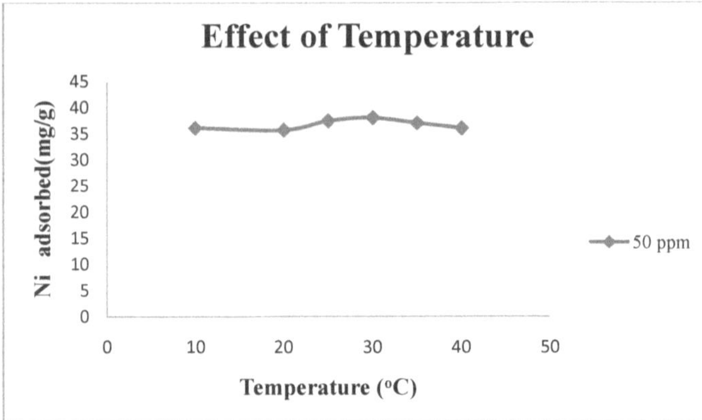

Fig4.88 Effect of Temperature on biosorption of Ni by *Oscillatoria sp.* (biomass dose=0.1 g, pH=5, initial Ni ion concentration = 50 mg/L; temperature = 25 ◦C; agitation speed = 120 rpm; contact time = 120 min).

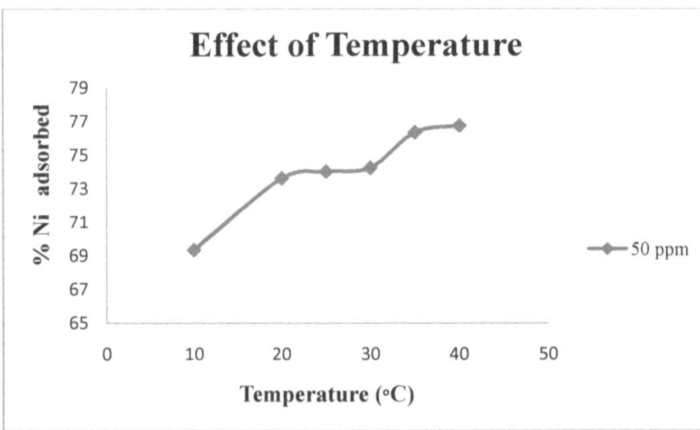

Fig4.89 Effect of Temperature on biosorption of Ni by *Spirulina sp.* (biomass dose=0.1 g, pH=5, initial Cu ion concentration = 50 mg/L; temperature = 25 ◦C; agitation speed = 120 rpm; contact time = 120 min).

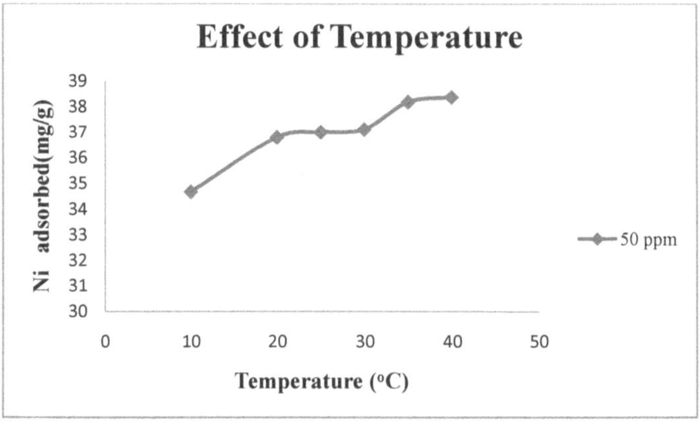

Fig4.90 Effect of Temperature on biosorption of Ni by *Spirulina sp.* (biomass dose=0.1 g, pH=5, initial Ni ion concentration = 50 mg/L; temperature = 25 ◦C; agitation speed = 120 rpm; contact time = 120 min).

4.16 Effect of temperature for Zinc

Results on the effect of temperature of Zn at initial metal ion concentration of 50 mg/L by *Chlorella pyrenoidosa, Oscillatoria sp.* and *Spirulina sp.* present in the Figures 4.91 to 4.96.

Maximum percentage of biosorption accursed at the initial concentration of 50 mg/L at the time of 120 minute at pH= 5 for Zn was 75.44%, 78.35% and 77.39% and metal ion uptake capacity was 37.72 mg/g, 39.06 mg/g and 38.69 mg/g at the temperature of 40 °C ,25 °C and 40 °C for *Chlorella pyrenoidosa, Oscillatoria sp.* and *Spirulina sp.* respectively.

The findings of *Chlorella pyrenoidosa* indicate that the sorption percentage increased with increase in temperature up to 40 ° C.

The findings of *Oscillatoria sp.* indicate that the temperature has no significant effect on the biosorption of Zn over tested 10-40 °C at pH= 5. Biosorption to be a passive energy independent process.

The findings of *Spirulina sp.* indicate that the sorption percentage increased with increase in temperature up to 40 ° C.

In effect of temperature and initial concentration of 50 mg/L, percentage of remove of Zn and metal ion uptake capacity *for the three biomass under study has been established as:*

Oscillatoria sp. > *Spirulina sp.* > *Chlorella pyrenoidosa*

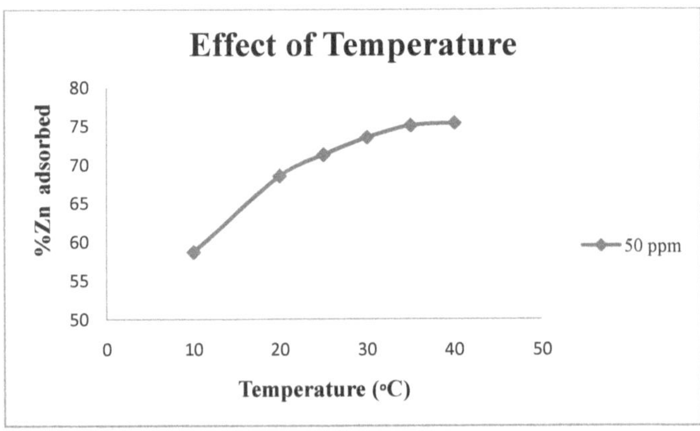

Fig4.91 Effect of Temperature on biosorption of Zn by *Chlorella pyrenoidosa* (biomass dose=0.1 g, pH=5, initial Zn ion concentration = 50 mg/L; temperature = 25 °C; agitation speed = 120 rpm; contact time = 120 min).

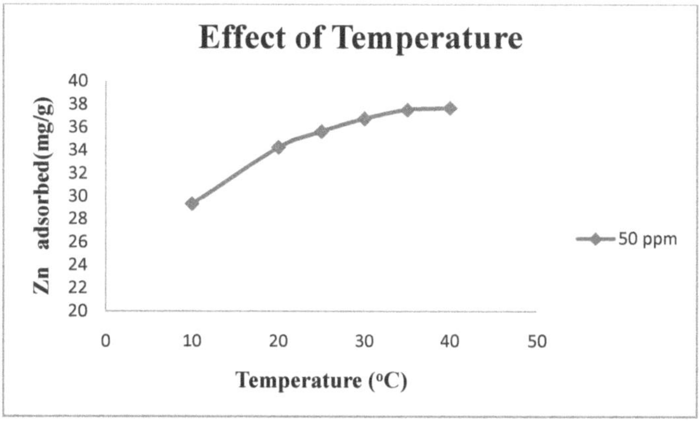

Fig4.92 Effect of Temperature on biosorption of Zn by *Chlorella pyrenoidosa* (biomass dose=0.1 g, pH=5, initial Zn ion concentration = 50 mg/L; temperature = 25 ◦C; agitation speed = 120 rpm; contact time = 120 min).

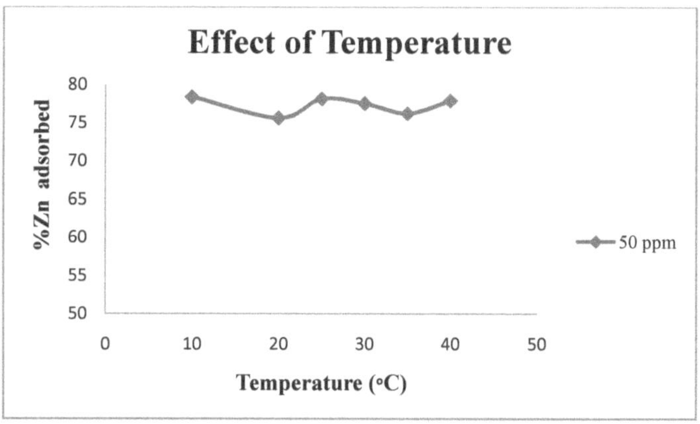

Fig4.93 Effect of Temperature on biosorption of Zn by *Oscillatoria sp.* (biomass dose=0.1 g, pH=5, initial Zn ion concentration = 50 mg/L; temperature = 25 ◦C; agitation speed = 120 rpm; contact time = 120 min).

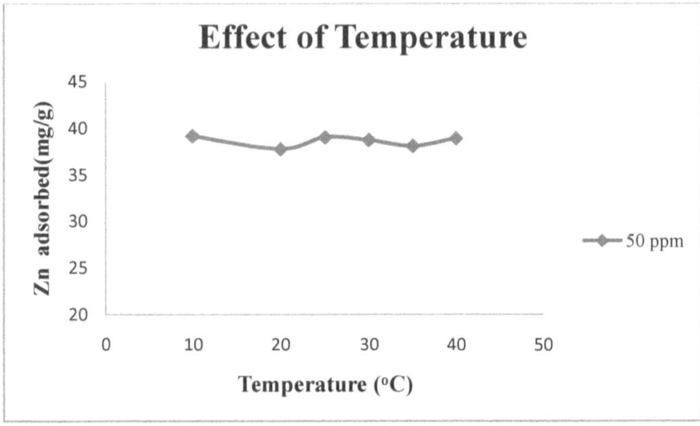

Fig4.94 Effect of Temperature on biosorption of Zn by *Oscillatoria sp.* (biomass dose=0.1 g, pH=5, initial Zn ion concentration = 50 mg/L; temperature = 25 ◦C; agitation speed = 120 rpm; contact time = 120 min).

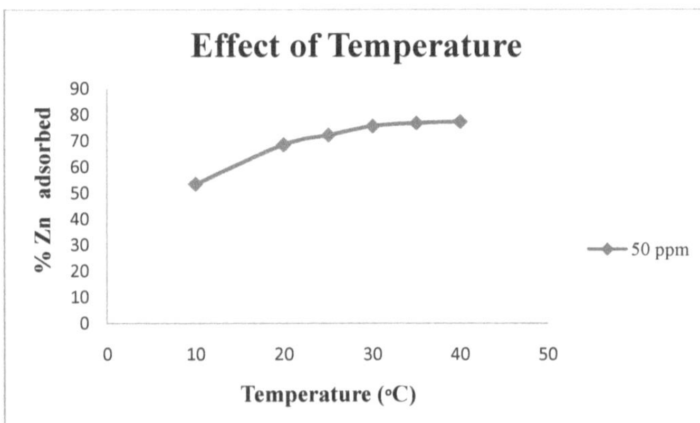

Fig4.95 Effect of Temperature on biosorption of Zn by *Spirulina sp.* (biomass dose=0.1 g, pH=5, initial Zn ion concentration = 50 mg/L; temperature = 25 ◦C; agitation speed = 120 rpm; contact time = 120 min).

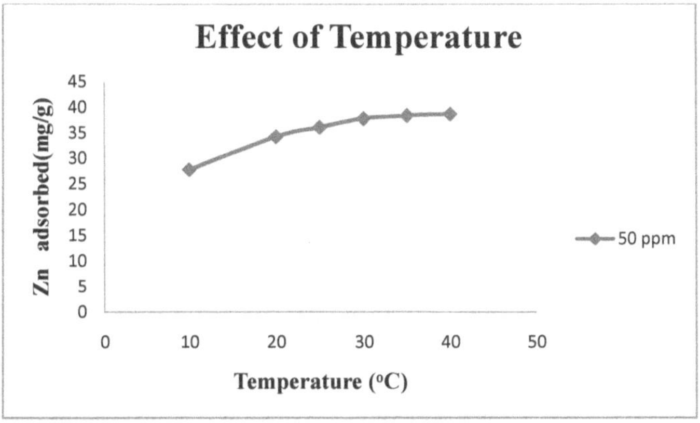

Fig4.96 Effect of Temperature on biosorption of Zn by *Spirulina sp.* (biomass dose=0.1 g, pH=5, initial Zn ion concentration = 50 mg/L; temperature = 25 ◦C; agitation speed = 120 rpm; contact time = 120 min).

4.17 Equilibrium isotherms for Chromium

Equilibrium isotherm study for biosorption of chromium by using *Chlorella pyrenoidosa, Oscillatoria sp.* and *Spirulina sp.* as shown in Figures 4.97 to 4.102.

Table 4.1 describes summaries of linear regression data for Langmuir and Freundlich isotherms for Chromium biosorption using *Chlorella pyrenoidosa, Oscillatoria sp.* and *Spirulina sp.* biomass. Langmuir and Freundlich constants were obtained from the linear equations of both models. As indicated in the Table 4.1 the coefficients of determination (R^2) of both models are close to 1. *The* constant of Freundlich (n) in Freundlich model for *the three biomasses under study* were bigger than 1 and showed the favorable adsorption.

Table 4.1 Parameters of isotherm models for Chromium

Biomass	Langmuir Parameters			Freundlich parameters		
	q_{max} (mg/g)	b(L/mg)	R^2	K_f	n	R^2

Chlorella pyrenoidosa	**125**	**0.049**	**0.809**	**7.82**	**1.53**	**0.991**
Oscillatoria sp.	**52.63**	**0.231**	**0.985**	**9.29**	**1.84**	**0.994**
Spirulina sp.	**90.91**	**0.045**	**0.953**	**5.65**	**1.56**	**0.997**

The maximum ions uptake capacity (q_{max}) in Langmuir model for *the three biomasses under study* for *has been established as:*

Chlorella pyrenoidosa > Spirulina sp. > Oscillatoria sp.

The equilibrium experimental results of Chromium ions have been fitted in the Langmuir and Freundlich models but coefficient of determination (R^2) Freundlich is bigger than of Langmuir and that means, Freundlich model is better model for our study. The value of 'R' is significant at 0.05 level according to the table given by Jog & Saptarshi (1980).

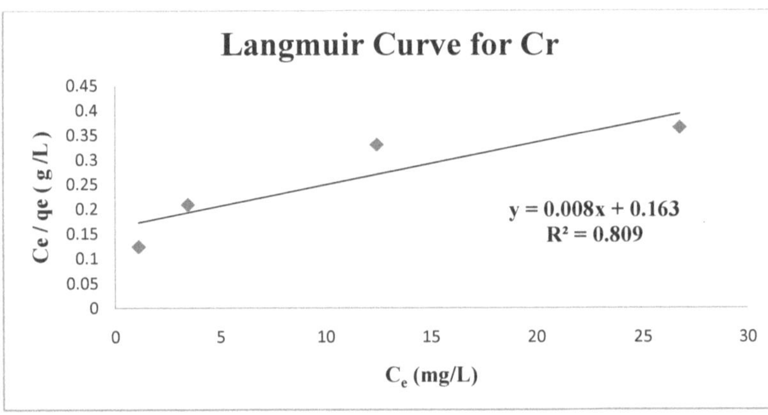

Fig.4.97 Langmuir adsorption isotherm for Cr by *Chlorella pyrenoidosa* (q_{max}= 125 mg/g , b= 0.049 L/mg).

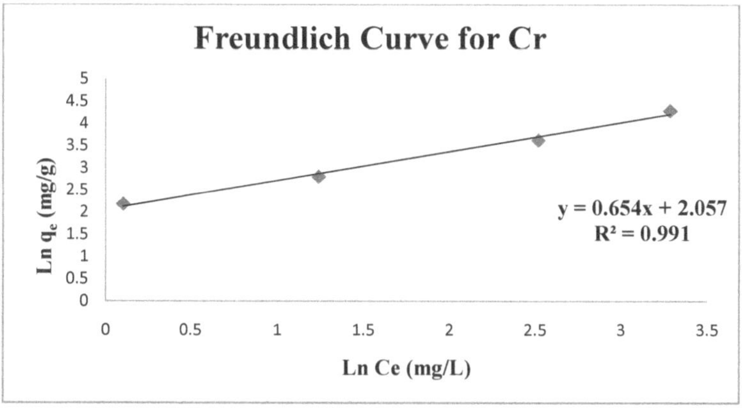

Fig.4.98 Freundlich adsorption isotherm for Cr by *Chlorella pyrenoidosa* (n=1.53, K= 7.82).

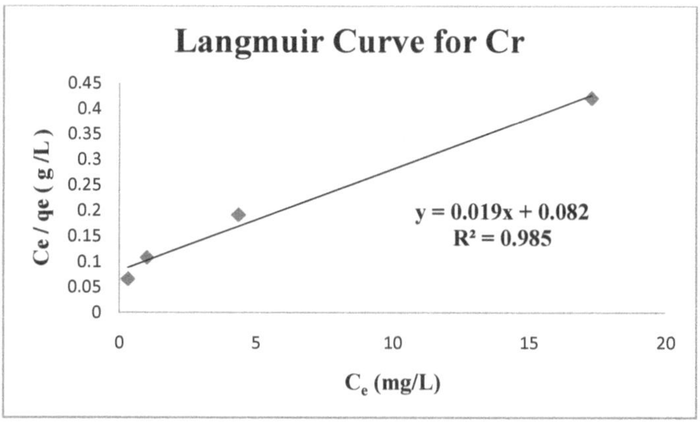

Fig.4.99 Langmuir adsorption isotherm for Cr by *Oscillatoria sp.* (q_{max}= 52.63 mg/g , b= 0.231 L/mg).

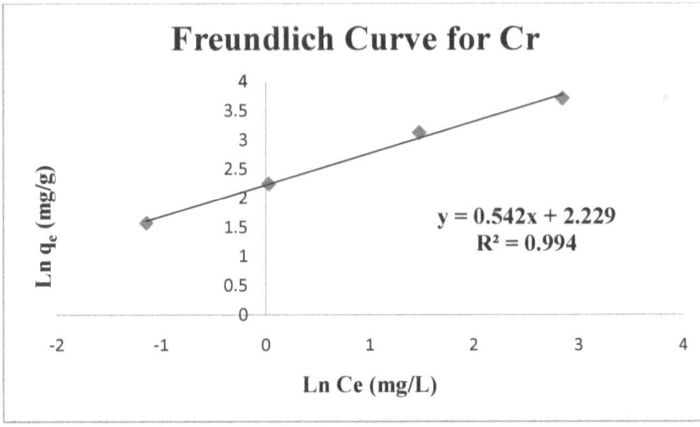

Fig.4.100 Freundlich adsorption isotherm for Cr by *Oscillatoria sp.* (n=1.84, K= 9.29).

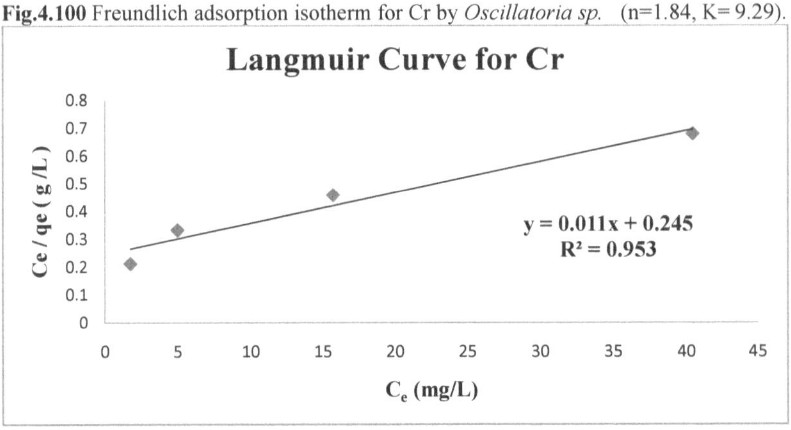

Fig.4.101 Langmuir adsorption isotherm for Cr by *Spirulina sp.* (q_{max}= 90.91 mg/g , b= 0.045 L/mg).

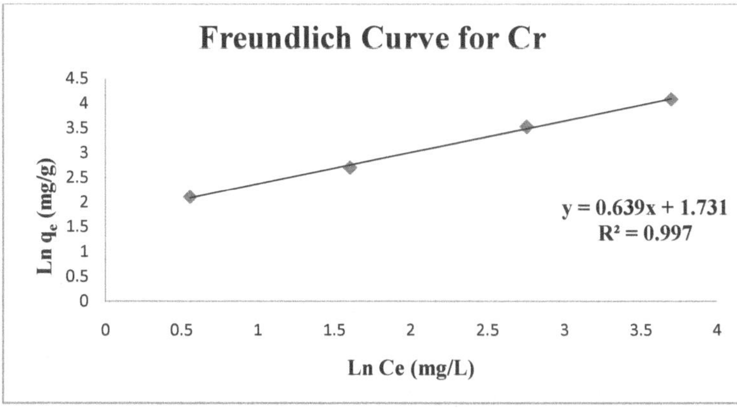

Fig.4.102 Freundlich adsorption isotherm for Cr by *Spirulina sp.* (n=1.56, K= 5.65).

4.18 Equilibrium isotherms for Copper

Study equilibrium isotherm for biosorption of copper by using *Chlorella pyrenoidosa, Oscillatoria sp.* and *Spirulina sp.* as shown in Figures 4.103 to 4.108.

Table4.2describes summaries of linear regression data for Langmuir and Freundlich isotherms for Copper biosorption using *Chlorella pyrenoidosa, Oscillatoria sp.* and *Spirulina sp.* biomass. Langmuir and Freundlich constant k were obtained from the linear equations of both models. As indicated in the Table 4.2, the coefficients of determination (R^2) of both models are close to 1. *The* constant of Freundlich (n) in Freundlich model for *the three biomasses under study* were bigger than 1 and showed the favorable adsorption.

Table 4.2 Parameters of isotherm models for Copper

Biomass	**Langmuir Parameters**			**Freundlich parameters**		
	q_{max} (mg/g)	**b**(L/mg)	R^2	K_f	n	R^2
Chlorella pyrenoidosa	**142.86**	**0.041**	**0.897**	**7.04**	**1.42**	**0.985**

Oscillatoria sp.	**500**	**0.021**	**0.790**	**10.62**	**1.13**	**0.998**
Spirulina sp.	**142.85**	**0.047**	**0.982**	**7.34**	**1.35**	**0.980**

The maximum ions uptake capacity (q_{max}) in Langmuir model for *the three biomasses under study* for *has been established as:*

Oscillatoria sp. > Chlorella pyrenoidosa > Spirulina sp.

The equilibrium experimental results of Copper ions have been fitted in the Langmuir and Freundlich models but coefficient of determination (R^2) Freundlich is bigger than of Langmuir and that means, Freundlich model is better model for our study. The value of 'R' is significant at 0.05 level according to the table given by Jog & Saptarshi (1980).

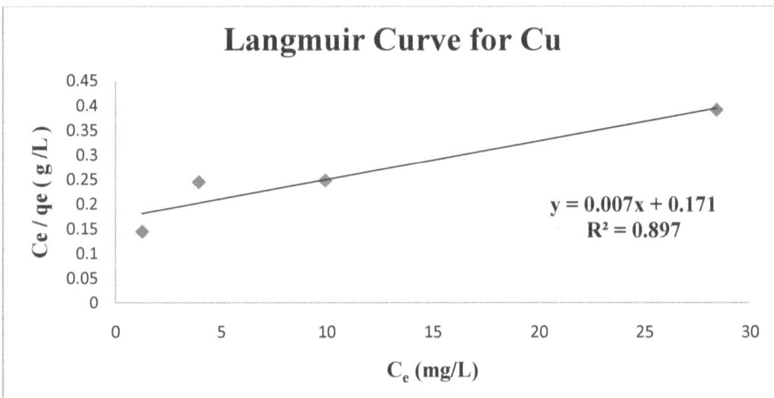

Fig.4.103 Langmuir adsorption isotherm for Cu by *Chlorella pyrenoidosa* (q_{max}= 142.86 mg/g , b= 0.041 L/mg).

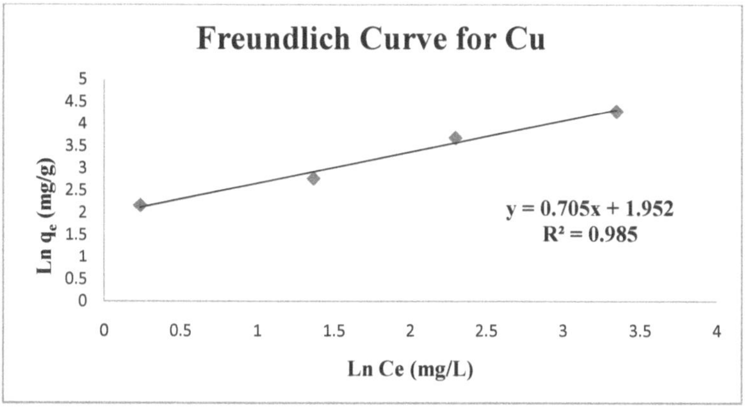

Fig.4.104 Freundlich adsorption isotherm for Cu by *Chlorella pyrenoidosa* (n=1.42, K= 7.04).

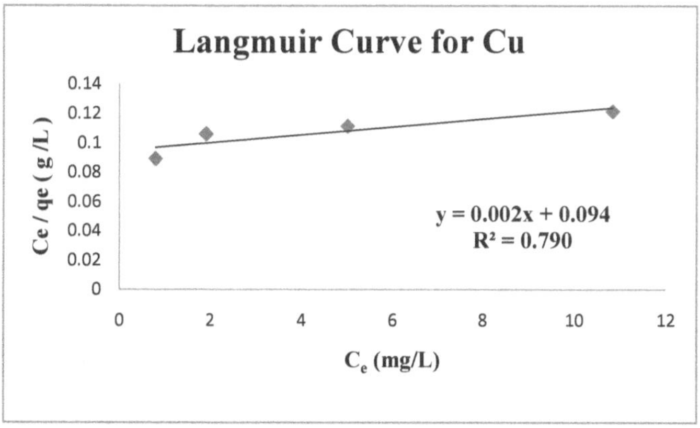

Fig.4.105 Langmuir adsorption isotherm for Cu by *Oscillatoria sp.* (q_{max}= 500 mg/g , b= 0.021 L/mg).

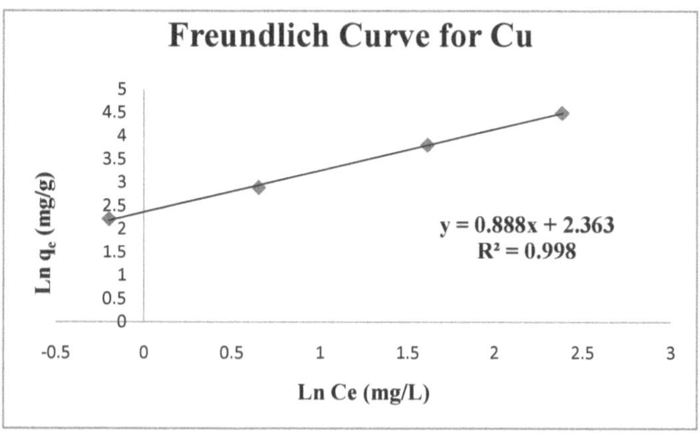

Fig.4.106 Freundlich adsorption isotherm for Cu by *Oscillatoria sp.* (n=1.13, K= 10.62).

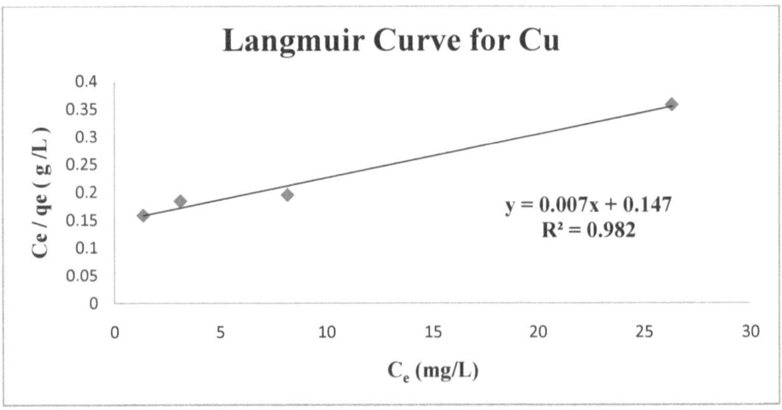

Fig.4.107 Langmuir adsorption isotherm for Cu by *Spirulina sp.* (q_{max}= 142.85 mg/g , b= 0.047L/mg).

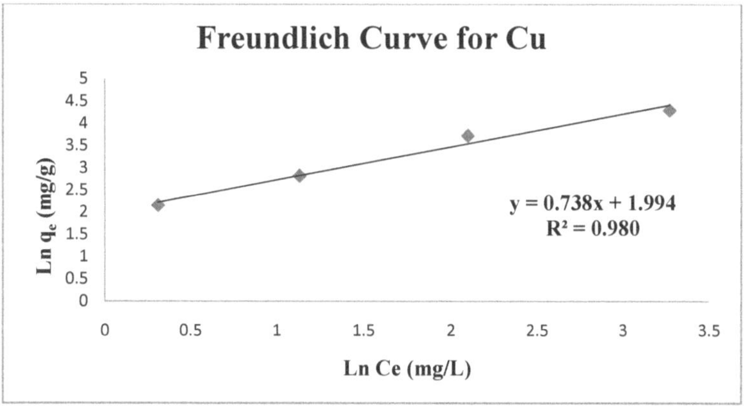

Fig.4.108 Freundlich adsorption isotherm for Cu by *Spirulina sp.* (n=1.35, K= 7.34).

4.19 Equilibrium isotherms for Nickel

Study of equilibrium isotherm for biosorption of nickel by using *Chlorella pyrenoidosa, Oscillatoria sp.* and *Spirulina sp.* as shown in Figures 4.109 to 4.114.

Table 4.3 describes summaries of linear regression data for Langmuir and Freundlich isotherms for Nickel biosorption using *Chlorella pyrenoidosa, Oscillatoria sp.* and *Spirulina sp.* biomass. Langmuir and Freundlich constant k were obtained from the linear equations of both models. As indicated in the Table 4.3, the coefficients of determination (R^2) of both models are close to 1. *The* constant of Freundlich (n) in Freundlich model for *the three biomasses under study* were bigger than 1 and showed the favorable adsorption.

Table 4.3 Parameters of isotherm models for Nickel

Biomass	Langmuir Parameters			Freundlich parameters		
	q_{max} (mg/g)	b(L/mg)	R^2	K_f	n	R^2
Chlorella pyrenoidosa	142.86	0.072	0.951	10.63	1.43	0.996

Oscillatoria sp.	**125**	**0.039**	**0.990**	**5.73**	**1.35**	**0.986**
Spirulina sp.	**111.11**	**0.035**	**0.961**	**5.05**	**1.41**	**0.999**

The maximum ions uptake capacity (q_{max}) in Langmuir model for *the three biomasses under study* for *has been established as:*

Chlorella pyrenoidosa > *Oscillatoria sp.* > *Spirulina sp.*

The equilibrium experimental results of Nickel ions have been fitted in the Langmuir and Freundlich models but coefficient of determination (R^2) Freundlich is bigger than of Langmuir and that means, Freundlich model is better model for our study. The value of 'R' is significant at 0.05 level according to the table given by Jog & Saptarshi (1980).

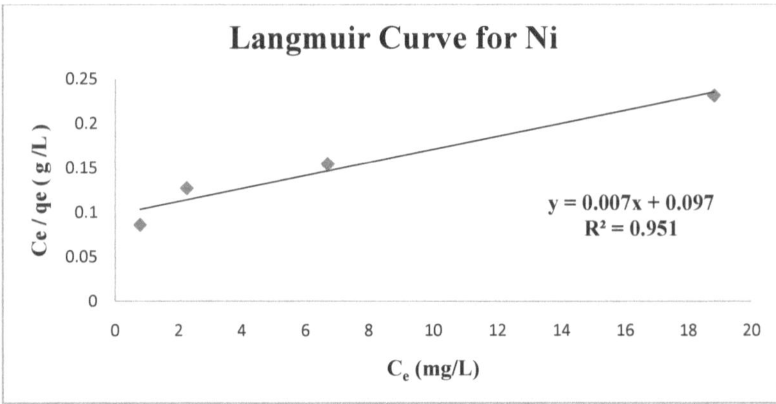

Fig.4.109 Langmuir adsorption isotherm for Ni by *Chlorella pyrenoidosa* (q_{max}= 142.86 mg/g , b= 0.072 L/mg).

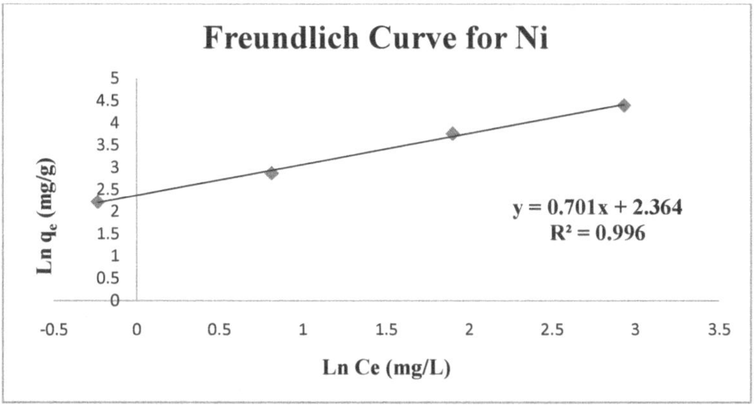

Fig.4.110 Freundlich adsorption isotherm for Ni by *Chlorella pyrenoidosa* (n=1.43, K= 10.63).

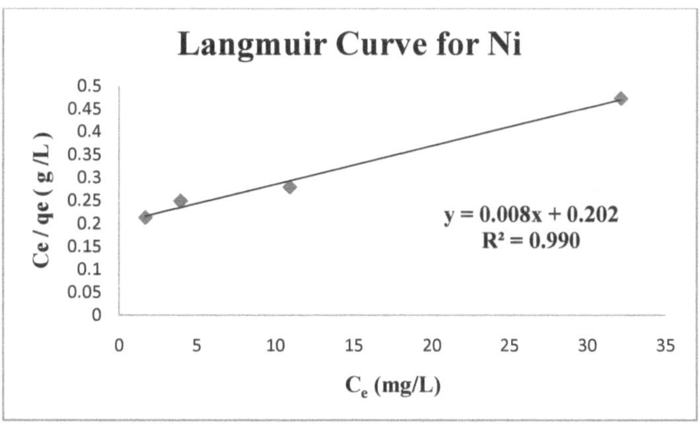

Fig.4.111 Langmuir adsorption isotherm for Ni by *Oscillatoria sp.* (q_{max}= 125 mg/g , b= 0.039).

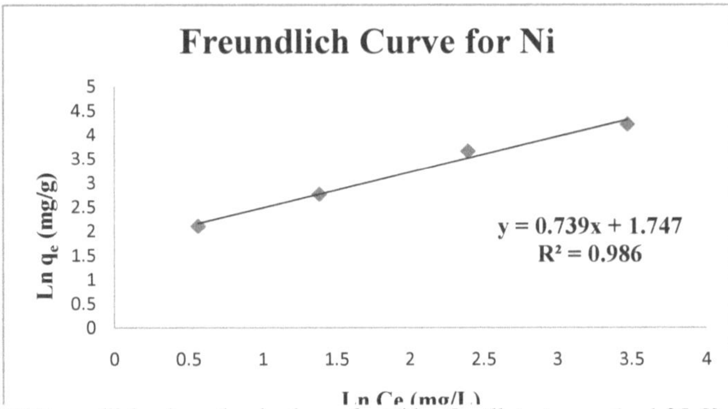

Fig.4.112 Freundlich adsorption isotherm for Ni by *Oscillatoria sp.* (n=1.35, K= 5.73).

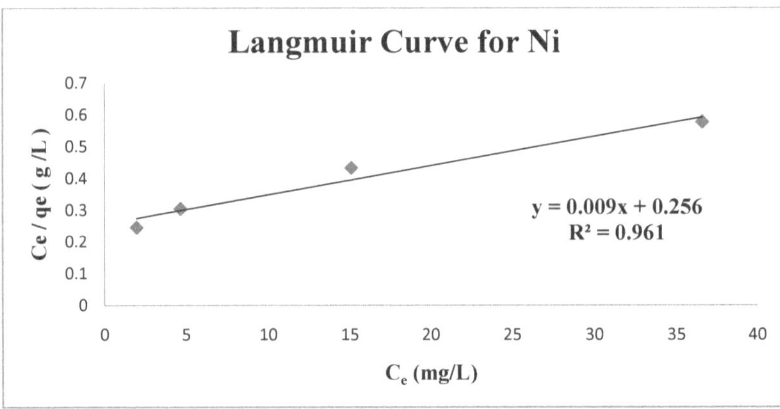

Fig.4.113 Langmuir adsorption isotherm for Ni by *Spirulina sp.* (q_{max}= 111.11 mg/g , b= 0.035 L/mg).

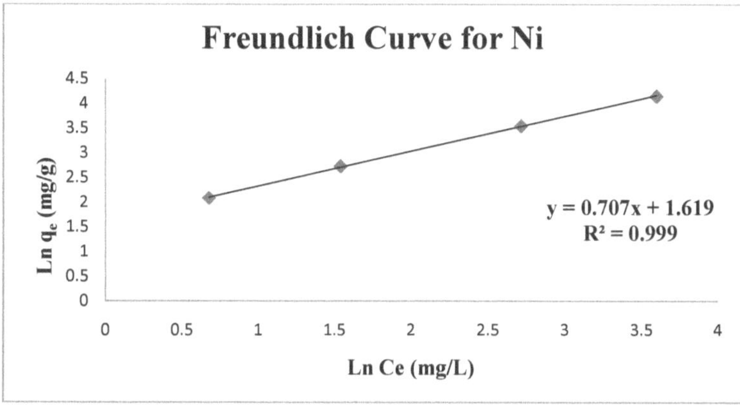

Fig.4.114 Freundlich adsorption isotherm for Ni by *Spirulina sp.* (n=1.41, K= 5.05).

4.20 Equilibrium isotherms for Zinc

Study of equilibrium isotherm for biosorption of zinc by using *Chlorella pyrenoidosa, Oscillatoria sp.* and *Spirulina sp.* as shown in Figures 4.115 to 4.120.

Table4.4 describes summaries of linear regression data for Langmuir and Freundlich isotherms for Zinc biosorption using *Chlorella pyrenoidosa, Oscillatoria sp.* and *Spirulina sp.* biomass. Langmuir and Freundlich constant k were obtained from the linear equations of both models. As indicated in the Table 4.4, the coefficients of determination (R^2) of both models are close to 1. The constant of Freundlich (n) in Freundlich model for the three biomasses under study were bigger than 1 and showed the favorable adsorption.

Table 4.4 Parameters of isotherm models for Zinc

Biomass	Langmuir Parameters			Freundlich parameters		
	q_{max} (mg/g)	b(L/mg)	R^2	K_f	n	R^2
Chlorella pyrenoidosa	101.11	0.036	0.918	5.38	1.44	0.999

Oscillatoria sp.	**125**	**0.048**	**0.939**	**7.20**	**1.44**	**0.998**
Spirulina sp.	**111.11**	**0.043**	**0.867**	**6.31**	**1.49**	**0.996**

The maximum ions uptake capacity (q_{max}) in Langmuir model for *the three biomasses under study* for *has been established as:*

Oscillatoria sp. > *Spirulina sp.* > *Chlorella pyrenoidosa*

The equilibrium experimental results of Zinc ions have been fitted in the Langmuir and Freundlich models but coefficient of determination (R^2) Freundlich is bigger than of Langmuir and that means, Freundlich model is better model for our study. The value of 'R' is significant at 0.05 level according to the table given by Jog & Saptarshi (1980).

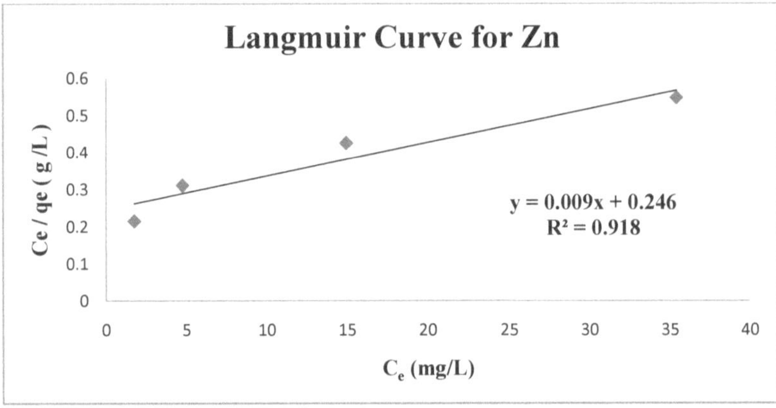

Fig.4.115 Langmuir adsorption isotherm for Zn by *Chlorella pyrenoidosa* (q_{max}= 101.11 mg/g , b= 0.036 L/mg).

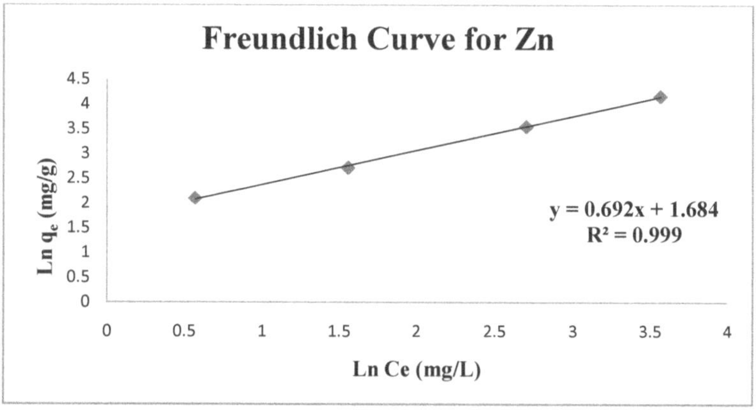

Fig.4.116 Freundlich adsorption isotherm for Zn by *Chlorella pyrenoidosa* (n=1.44, K= 5.38).

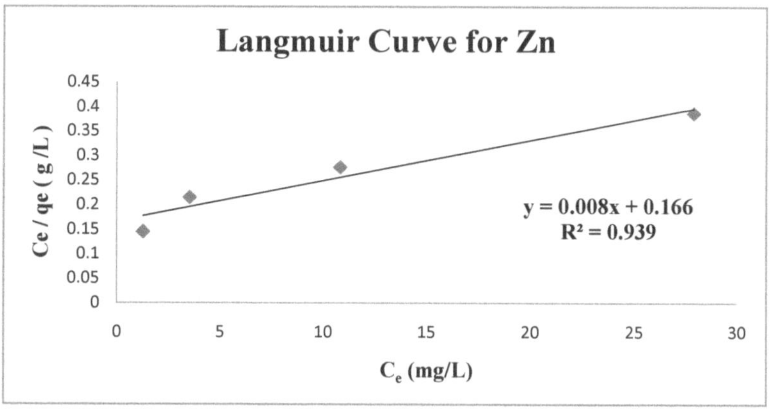

Fig.4.117 Langmuir adsorption isotherm for Zn by *Oscillatoria sp.* (q_{max}= 125 mg/g , b= 0.048).

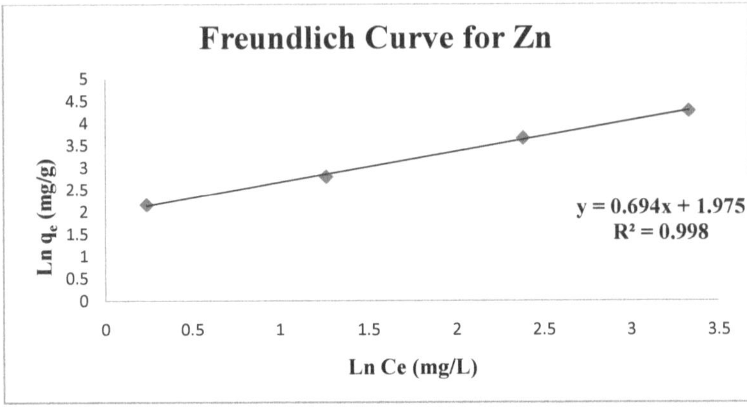

Fig.4.118 Freundlich adsorption isotherm for Zn by *Oscillatoria sp.* (n=1.44, K= 7.20).

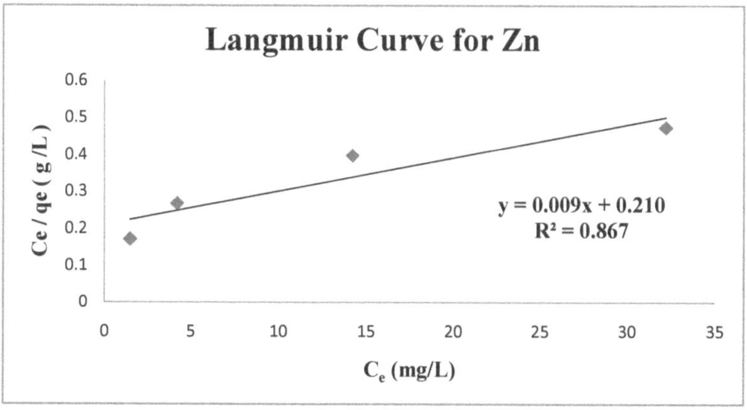

Fig.4.119 Langmuir adsorption isotherm for Zn by *Spirulina sp.* (q_{max}= 111.11 mg/g , b= 0.043 L/mg).

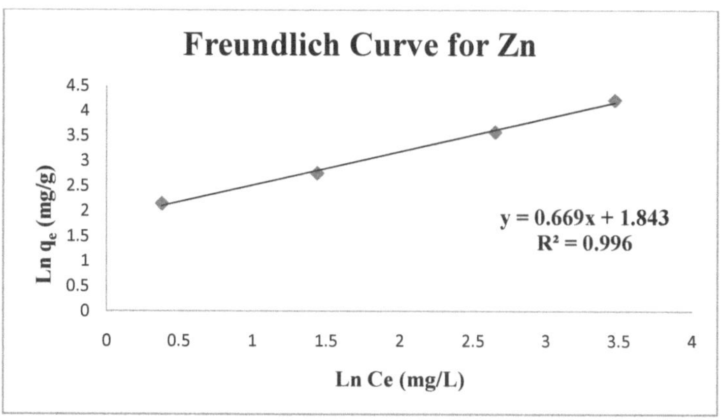

Fig.4.120 Freundlich adsorption isotherm for Zn by *Spirulina sp.* (n=1.49, K= 6.31).

4.21 Equilibrium parameter R_L

The essential characteristics of a Langmuir isotherm can be expressed in terms of a dimensionless constant separation factor or equilibrium parameter R_L, which is defined by

$$R_L = 1/1+bC_o$$

Where C_o is the initial adsorbate concentration (mg/L) and b is the Langmuir constant (L/mg). The parameter indicates the shape of the isotherm as follows (Table4.5):

Table 4.5: Type of isotherm for various R_L

R_L	$R_L>1$	$R_L=1$	$0< R_L<1$	$R_L = 0$
Type of isotherm	Un favorable	Linear	Favorable	Irreversible

The R_L values at different initial adsorbate concentrations indicate favorable adsorption for all the adsorbents and adsorbates studied.

4.22 Kinetic modeling for Chromium

Study kinetic model for biosorption of chromium by using *Chlorella pyrenoidosa, Oscillatoria sp.* and *Spirulina sp.* as shown in Figures 4.121 to 4.123and also parameter of kinetic model for Chromium of *Chlorella pyrenoidosa, Oscillatoria sp.* and *Spirulina sp* were summarized on the Tables 4.6 to 4.8.

Results showed that *the* coefficient of determination (R^2) and K_2 HO'S Model for *the different metal ion concentration under study* for *has been established as:*

10 mg/L > 20 mg/L > 50 mg/L > 100 mg/L

With increase the initial concentration coefficient of determination (R^2) and K_2 decreased. The value of 'R' is significant at 0.05 level according to the table given by Jog & Saptarshi (1980).

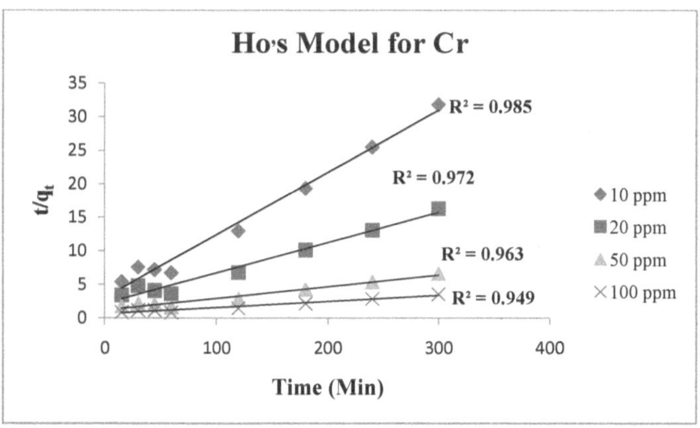

Fig.4.121 Plot of HO'S Model (Pseudo Second Order Rate) for Cr by *Chlorella pyrenoidosa*

Table 4.6 Parameter of kinetic model for Chromium of *Chlorella pyrenoidosa*

Concentration of Cr	Equation	R^2	K_2
10 mg/L	y = 0.093x + 3.054	0.985	2.83×10^{-3}
20 mg/L	y = 0.045x + 2.189	0.972	9.25×10^{-4}
50 mg/L	y = 0.017x + 1.182	0.963	2.44×10^{-4}
100 mg/L	y = 0.009x + 0.637	0.949	1.27×10^{-4}

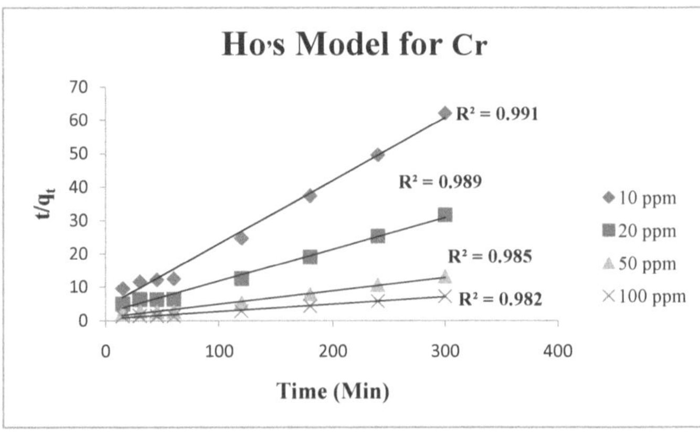

Fig.4.122 Plot of HO'S Model (Pseudo Second Order Rate) for Cr by *Oscillatoria sp.*

Table 4.7 Parameter of kinetic model for Chromium of *Oscillatoria sp.*

Concentration of Cr	Equation	R^2	K_2
10 mg/L	y = 0.188x + 4.154	0.991	8.51×10^{-3}
20 mg/L	y = 0.095x + 2.322	0.989	3.88×10^{-3}
50 mg/L	y = 0.039x + 1.055	0.985	1.44×10^{-3}
100 mg/L	y = 0.021x + 0.608	0.982	7.25×10^{-4}

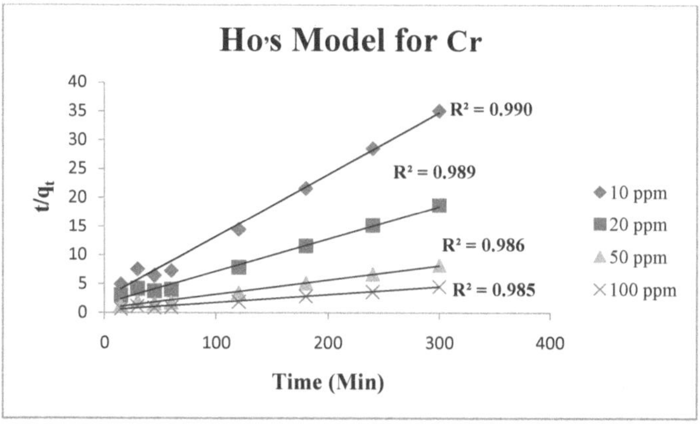

Fig.4.123 Plot of HO'S Model (Pseudo Second Order Rate) for Cr by *Spirulina sp.*

Table 4.8 Parameter of kinetic model for Chromium of *Spirulina sp.*

Concentration of Cr	Equation	R^2	K_2
10 mg/L	**y = 0.107x + 2.477**	**0.990**	**4.62×10^{-3}**
20 mg/L	**y = 0.055x + 1.612**	**0.989**	**1.87×10^{-3}**
50 mg/L	**y = 0.024x + 0.712**	**0.986**	**8.09×10^{-4}**
100 mg/L	**y = 0.013x + 0.424**	**0.985**	**3.98×10^{-4}**

4.23 Kinetic modeling for Copper

Study kinetic model for biosorption of copper by using *Chlorella pyrenoidosa, Oscillatoria sp.* and *Spirulina sp.* as shown in Figures 4.124 to 4.126 and also parameter of kinetic model for Chromium of *Chlorella pyrenoidosa, Oscillatoria sp.* and *Spirulina sp* were summarized on the Tables 4.9 to 4.11.Results showed that *the* coefficient of determination (R^2) and K_2 HO'S Model for *the different metal ion concentration under study* for *has been established as:* *10 mg/L* > 20 mg/L > 50 mg/L > 100 mg/L

With increase the initial concentration coefficient of determination (R^2) and K_2 decreased. The value of 'R' is significant at 0.05 level according to the table given by Jog & Saptarshi (1980).

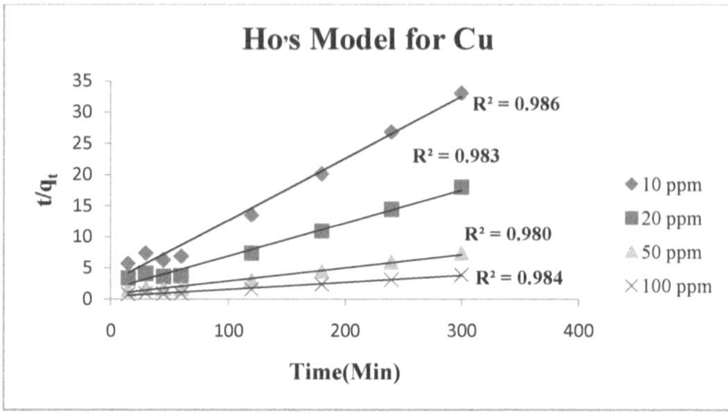

Fig.4.124 Plot of HO'S Model (Pseudo Second Order Rate) for Cu by *Chlorella pyrenoidosa*

Table 4.9 Parameter of kinetic model for Copper of *Chlorella pyrenoidosa*

Concentration of Cu	Equation	R^2	K_2
10 mg/L	y = 0.099x + 2.720	0.986	3.60×10^{-3}
20 mg/L	y = 0.053x + 1.636	0.983	1.72×10^{-3}
50 mg/L	y = 0.021x + 0.793	0.980	5.56×10^{-4}
100 mg/L	y = 0.011x + 0.483	0.984	2.50×10^{-4}

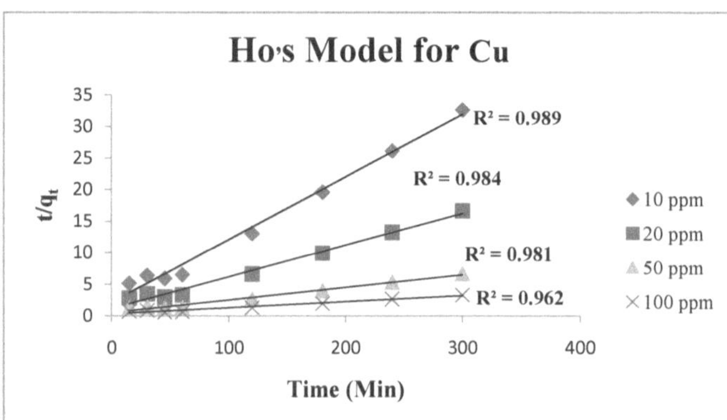

Fig.4.125 Plot of HO'S Model (Pseudo Second Order Rate) for Cu by *Oscillatoria sp.*

Table 4.10 Parameter of kinetic model for Copper of *Oscillatoria sp.*

Concentration of Cu	Equation	R^2	K_2
10 mg/L	y = 0.099x + 2.159	0.989	4.54×10^{-3}
20 mg/L	y = 0.050x + 1.197	0.984	2.09×10^{-3}
50 mg/L	y = 0.019x + 0.544	0.981	6.63×10^{-4}
100 mg/L	y = 0.009x + 0.372	0.962	2.18×10^{-4}

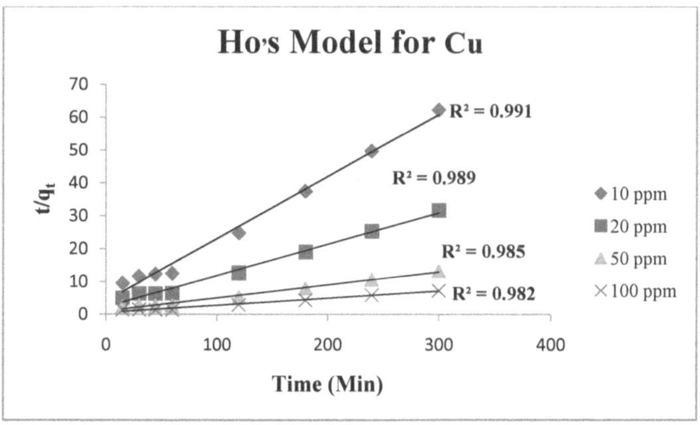

Fig.4.126 Plot of HO'S Model (Pseudo Second Order Rate) for Cu by *Spirulina sp.*

Table 4.11 Parameter of kinetic model for Copper of *Spirulina sp.*

Concentration of Cu	Equation	R^2	K_2
10 mg/L	y = 0.188x + 4.154	0.991	8.51×10^{-3}
20 mg/L	y = 0.095x + 2.322	0.989	3.89×10^{-3}
50 mg/L	y = 0.039x + 1.055	0.985	1.44×10^{-3}
100 mg/L	y = 0.021x + 0.608	0.982	7.25×10^{-4}

4.24 Kinetic modeling for Nickel

Study kinetic model for biosorption of nickel by using *Chlorella pyrenoidosa, Oscillatoria sp.* and *Spirulina sp.* as shown in Figures 4.127 to 4.129and also parameter of kinetic model for Chromium of *Chlorella pyrenoidosa, Oscillatoria sp.* and *Spirulina sp* were summarized on the Tables 4.12 to 4.14.Results showed that *the* coefficient of determination (R^2) and K_2 HO'S Model for *the different metal ion concentration under study* for *has been established as:* *10 mg/L* > 20 mg/L > 50 mg/L > 100 mg/L

With increase the initial concentration coefficient of determination (R^2) and K_2 decreased. The value of 'R' is significant at 0.05 level according to the table given by Jog & Saptarshi (1980).

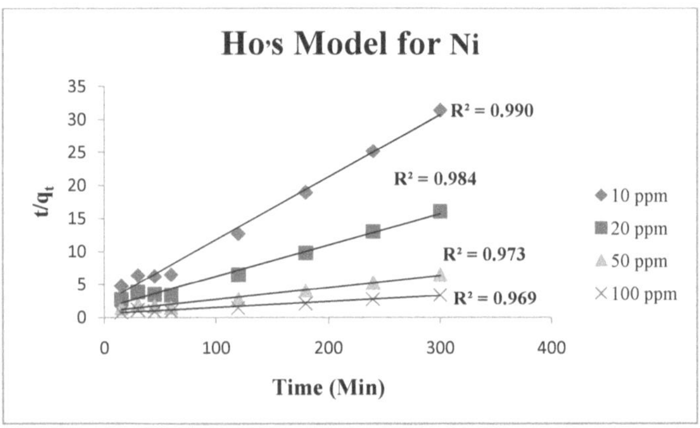

Fig.4.127 Plot of HO'S Model (Pseudo Second Order Rate) for Ni by *Chlorella pyrenoidosa*

Table 4.12 Parameter of kinetic model for Nickel of *Chlorella pyrenoidosa*

Concentration of Ni	Equation	R^2	K_2
10 mg/L	**y = 0.094x + 2.320**	**0.990**	$\mathbf{3.38 \times 10^{-3}}$
20 mg/L	**y = 0.047x + 1.526**	**0.984**	$\mathbf{1.44 \times 10^{-3}}$
50 mg/L	**y = 0.018x + 0.930**	**0.973**	$\mathbf{3.48 \times 10^{-4}}$
100 mg/L	**y = 0.009x + 0.602**	**0.969**	$\mathbf{1.34 \times 10^{-4}}$

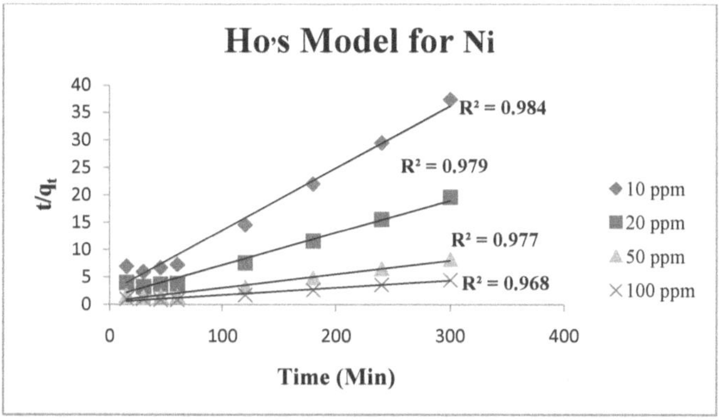

Fig.4.128 Plot of HO'S Model (Pseudo Second Order Rate) for Ni by *Oscillatoria sp.*

Table 4.13 Parameter of kinetic model for Nickel of *Oscillatoria sp.*

Concentration of Ni	Equation	R^2	K_2
10 mg/L	y = 0.113x + 2.320	0.984	5.50×10^{-3}
20 mg/L	y = 0.058x + 1.341	0.979	2.51×10^{-3}
50 mg/L	y = 0.013x + 0.451	0.977	9.74×10^{-4}
100 mg/L	y = 0.013x + 0.451	0.968	3.74×10^{-4}

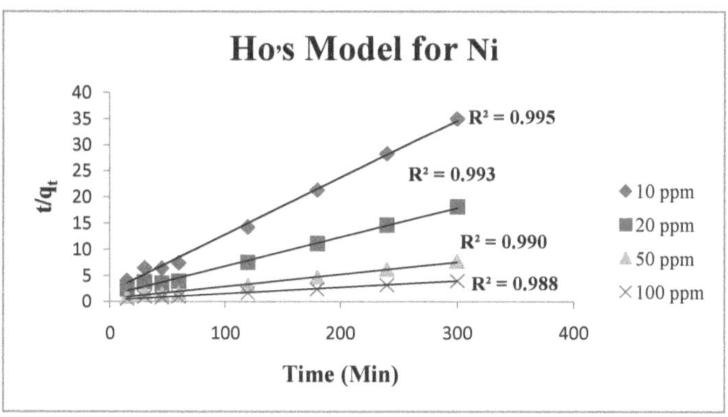

Fig.4.129 Plot of HO'S Model (Pseudo Second Order Rate) for Ni by *Spirulina sp.*

Table 4.14 Parameter of kinetic model for Nickel of *Spirulina sp.*

Concentration of Ni	Equation	R^2	K_2
10 mg/L	y = 0.108x + 1.965	0.995	5.93×10^{-3}
20 mg/L	y = 0.055x + 1.296	0.993	2.33×10^{-3}
50 mg/L	y = 0.011x + 0.377	0.990	8.59×10^{-4}
100 mg/L	y = 0.011x + 0.377	0.988	3.21×10^{-4}

4.25 Kinetic modeling for Zinc

Study kinetic model for biosorption of zinc by using *Chlorella pyrenoidosa, Oscillatoria sp.* and *Spirulina sp.* as shown in Figures 4.130 to 4.132and also parameter of kinetic model for Chromium of *Chlorella pyrenoidosa, Oscillatoria sp.* and *Spirulina sp* were summarized on the Tables 4.15 To 4.17.Results showed that *the* coefficient of determination (R^2) and K_2 HO'S Model for *the different metal ion concentration under study* for *has been established as:* *10 mg/L* > 20 mg/L > 50 mg/L > 100 mg/L

With increase the initial concentration coefficient of determination (R^2) and K_2 decreased. The value of 'R' is significant at 0.05 level according to the table given by Jog & Saptarshi (1980).

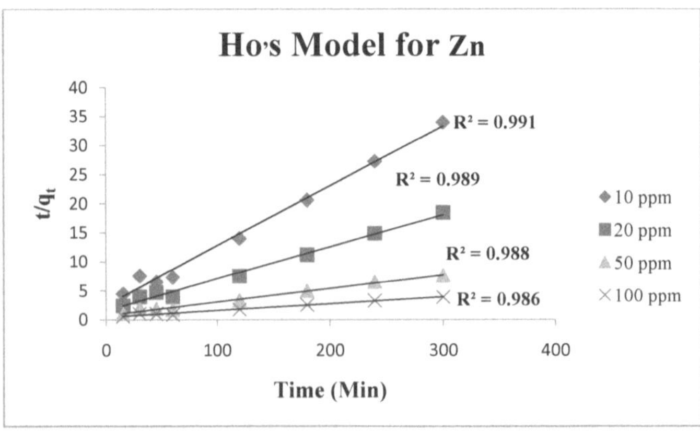

Fig.4.130 Plot of HO'S Model (Pseudo Second Order Rate) for Zn by *Chlorella pyrenoidosa*

Table 4.15 Parameter of kinetic model for Zinc of *Chlorella pyrenoidosa*

Concentration of Zn	Equation	R^2	K_2
10 mg/L	y = 0.103x + 2.468	0.991	4.30×10^{-3}
20 mg/L	y = 0.055x + 1.577	0.989	1.92×10^{-3}
50 mg/L	y = 0.023x + 0.762	0.988	9.94×10^{-4}
100 mg/L	y = 0.011x + 0.479	0.986	2.53×10^{-4}

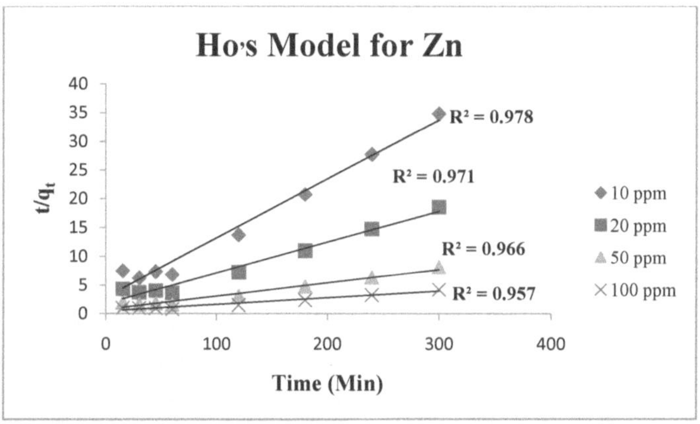

Fig.4.131 Plot of HO'S Model (Pseudo Second Order Rate) for Zn by *Oscillatoria sp.*

Table 4.16 Parameter of kinetic model for Zinc of *Oscillatoria sp.*

Concentration of Zn	Equation	R^2	K_2
10 mg/L	y = 0.102x + 2.972	0.978	3.50×10^{-3}
20 mg/L	y = 0.053x + 1.811	0.971	1.55×10^{-3}
50 mg/L	y = 0.023x + 0.756	0.966	6.99×10^{-4}
100 mg/L	y = 0.011x + 0.449	0.957	2.69×10^{-4}

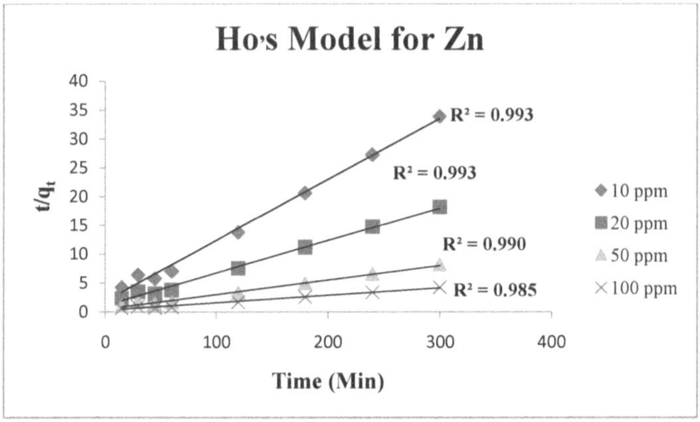

Fig.4.132 Plot of HO'S Model (Pseudo Second Order Rate) for Zn by *Spirulina sp.*

Table 4.17 Parameter of kinetic model for Zinc of *Spirulina sp.*

Concentration of Zn	**Equation**	R^2	K_2
10 mg/L	**y = 0.105x + 1.846**	**0.993**	5.97×10^{-3}
20 mg/L	**y = 0.056x + 1.098**	**0.993**	2.86×10^{-3}
50 mg/L	**y = 0.025x + 0.512**	**0.990**	1.22×10^{-3}
100 mg/L	**y = 0.012x + 0.339**	**0.985**	4.25×10^{-4}

4.26 Thermodynamic studies

Results on the Plots of Van' t Hoff equation for the estimation of thermodynamic parameters by *Chlorella pyrenoidosa, Oscillatoria sp.* and *Spirulina sp.* present in the Figures 4.133 to 4.144.

The calculated values of thermodynamic parameters are reported in Table 4.118 to 4.20 A positive value of ΔH° indicated the endothermic nature of the process. A negative value of the free energy (ΔG°) indicated the spontaneous nature of the adsorption process. It was also noted that the change in free energy, increases with increase in temperature, which exhibits an increase in adsorption with rise in temperature. This could be possibly because of activation of more sites on the surface of biomasses with increase in temperature. A positive value of ΔS° showed increased randomness at solid–liquid interface during the adsorption of heavy metals, it also suggests some structural changes in the adsorbate and the adsorbent.

The coefficients of determination (R^2) of *pyrenoidosa* and *Spirulina sp.* indicate that the temperature has significant effect on the biosorption of heavy metals. The value of 'R' for *Chlorella pyrenoidosa* and *Spirulina sp.* is significant at 0.05 level according to the table given by Jog & Saptarshi (1980).

The coefficients of determination (R^2) of *Oscillatoria sp.* indicate that the temperature has no significant effect on the biosorption of heavy metals. The value of 'R' is not significant at 0.05 level according to the table given by Jog & Saptarshi (1980).

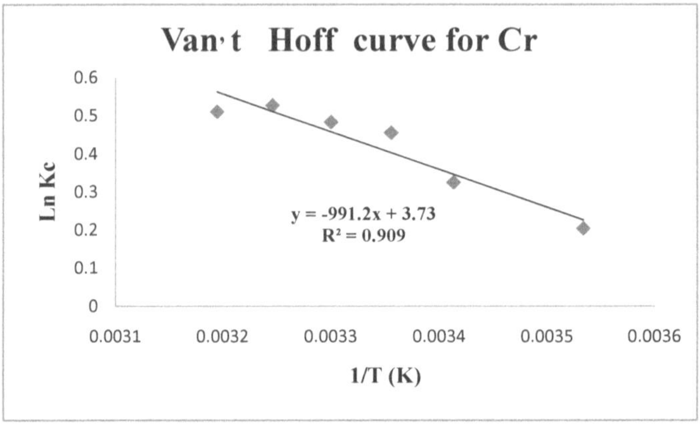

Fig. 4.133 Plot of Van' t Hoff equation for the estimation of thermodynamic parameters for biosorption of Cr by *Chlorella pyrenoidosa.*

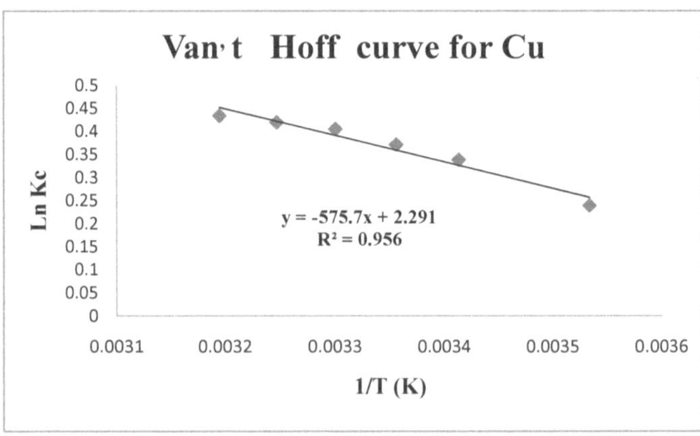

Fig. 4.134 Plot of Van' t Hoff equation for the estimation of thermodynamic parameters for biosorption of Cu by *Chlorella pyrenoidosa.*

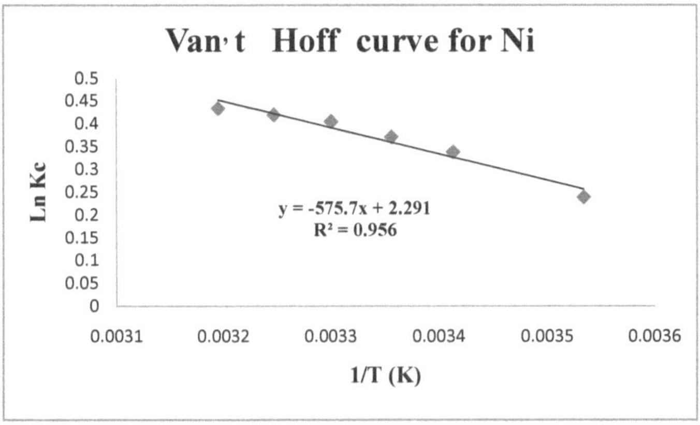

Fig. 4.135 Plot of Van' t Hoff equation for the estimation of thermodynamic parameters for biosorption of Ni by *Chlorella pyrenoidosa*.

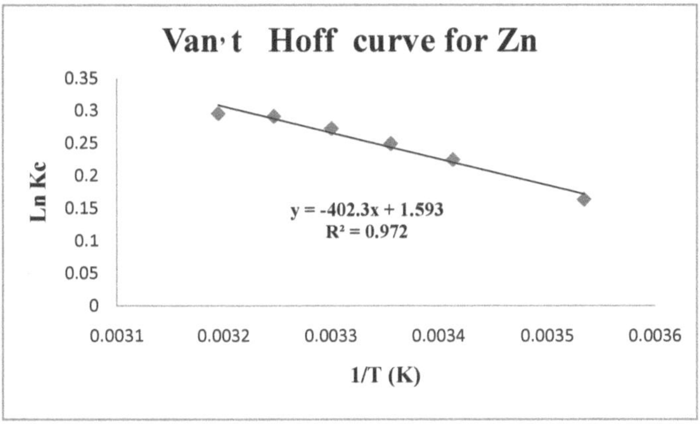

Fig. 4.136 Plot of Van' t Hoff equation for the estimation of thermodynamic parameters for biosorption of Zn by *Chlorella pyrenoidosa*.

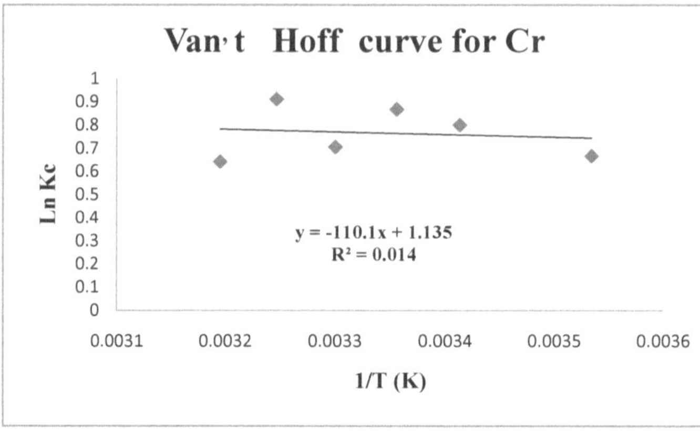

Fig. 4.137 Plot of Van' t Hoff equation for the estimation of thermodynamic parameters for biosorption of Cr by *Oscillatoria sp.*

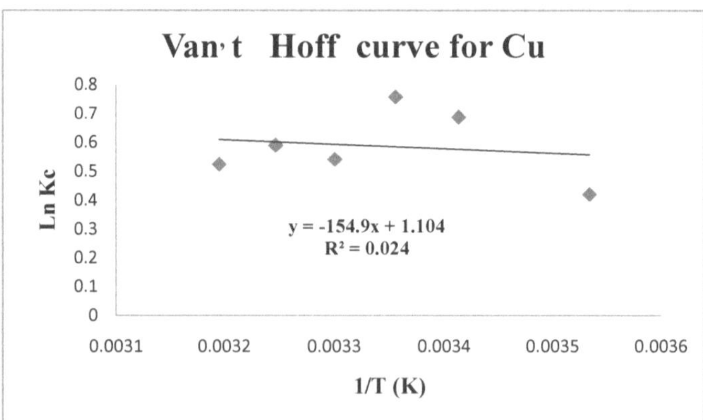

Fig. 4.138 Plot of Van' t Hoff equation for the estimation of thermodynamic parameters for biosorption of Cu by *Oscillatoria sp.*

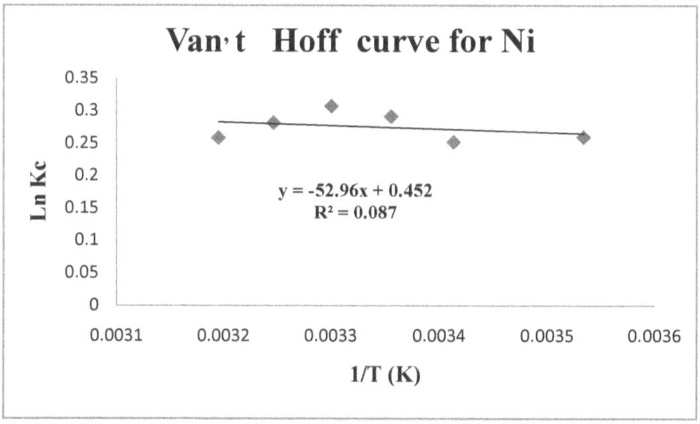

Fig. 4.139 Plot of Van't Hoff equation for the estimation of thermodynamic parameters for biosorption of Ni by *Oscillatoria sp.*

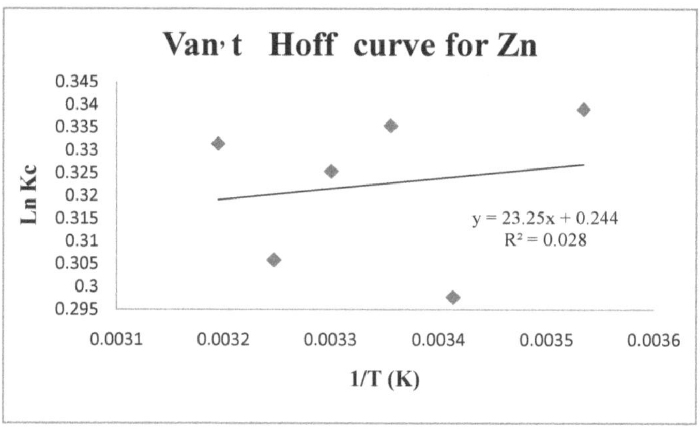

Fig. 4.140 Plot of Van't Hoff equation for the estimation of thermodynamic parameters for biosorption of Zn by *Oscillatoria sp.*

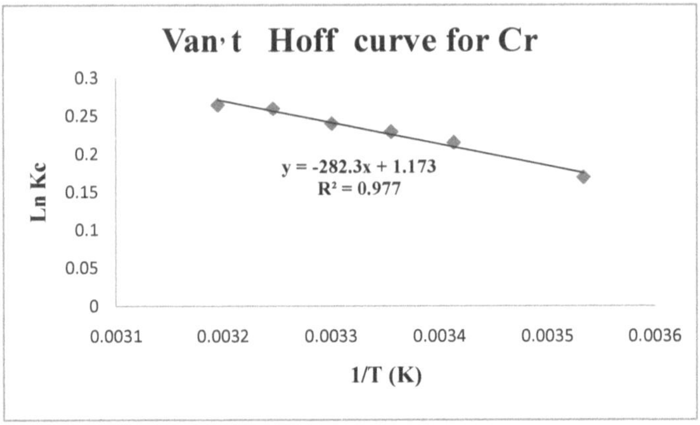

Fig. 4.141 Plot of Van' t Hoff equation for the estimation of thermodynamic parameters for biosorption of Cr by *Spirulina sp.*

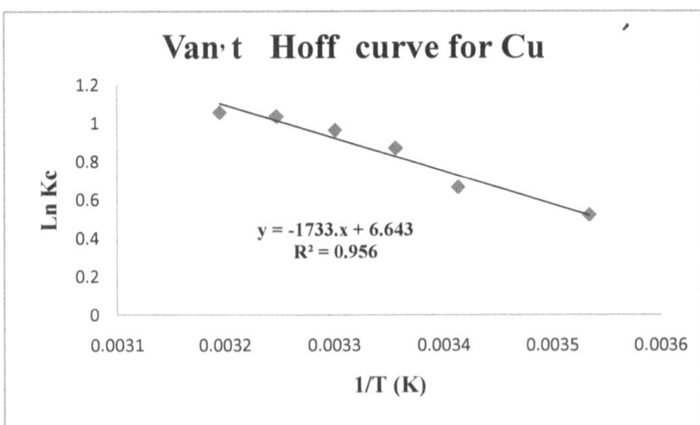

Fig. 4.142 Plot of Van' t Hoff equation for the estimation of thermodynamic parameters for biosorption of Cu by *Spirulina sp.*

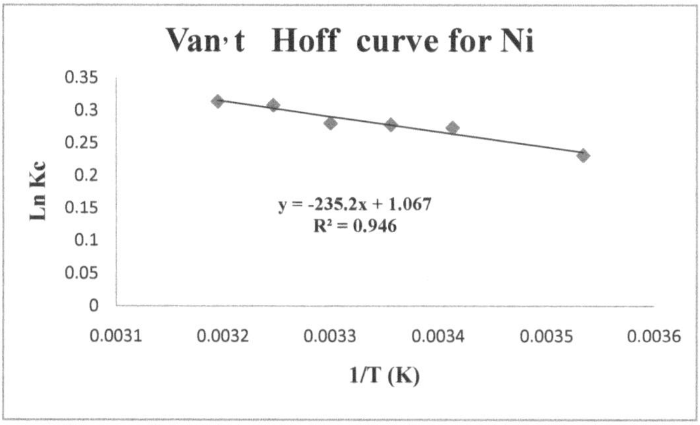

Fig. 4.143 Plot of Van' t Hoff equation for the estimation of thermodynamic parameters for biosorption of Ni by *Spirulina sp.*

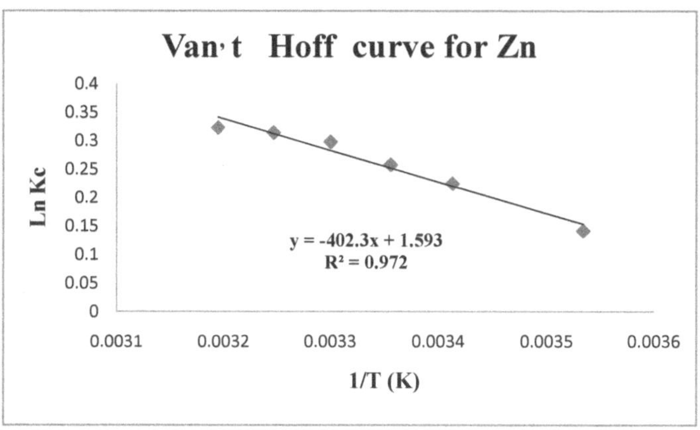

Fig. 4.144 Plot of Van' t Hoff equation for the estimation of thermodynamic parameters for biosorption of Zn by *Spirulina sp.*

Table 4.18 Thermodynamic parameters for the adsorption of heavy metals by *Chlorella*

Heavy Metals	ΔH° (J mol^{-1}K^{-1})	ΔS° (J K^{-1}mol^{-1})	-Δ°G (J.mol^{-1}K^{-1})					
			283 K	293 K	298 K	303 K	308 K	313 K
Cr	8240.83	31.01	17.06	54.21	94.86	120.81	153.60	170.05
Cu	4786.37	19.05	19.93	56.37	77.02	101.22	122.22	144.33
Ni	12063.61	44.73	19.07	70.86	116.89	147.84	194.24	238.90
Zn	3344.72	13.24	13.61	37.39	51.84	68.01	84.74	98.31

Table 4.19 Thermodynamic parameters for the adsorption of heavy metals by *Oscillatoria sp.*

Heavy Metals	ΔH° (J mol^{-1}K^{-1})	ΔS° (J K^{-1}mol^{-1})	-Δ°G (J.mol^{-1}K^{-1})					
			283 K	293 K	298 K	303 K	308 K	313 K
Cr	915.37	9.44	55.59	133.45	180.87	176.42	265.61	214.18
Cu	1287.84	9.18	34.92	114.53	157.59	135.02	171.64	174.27
Ni	440.31	3.76	21.59	41.89	60.58	76.65	81.93	86.07
Zn	164.78	3.27	28.18	49.51	69.70	81.16	98.93	110.22

Table 4.20 Thermodynamic parameters for the adsorption of heavy metals by *Spirulina sp.* .

Heavy Metals	ΔH° (J mol^{-1}K^{-1})	ΔS° (J K^{-1}mol^{-1})	-Δ°G (J.mol^{-1}K^{-1})					
			283 K	293 K	298 K	303 K	308 K	313 K
Cr	2347.04	9.75	14.10	35.83	47.69	59.92	75.68	88.04
Cu	14415.09	55.23	43.25	110.79	180.87	240.58	301.62	351.49
Ni	1955.45	8.87	19.26	45.53	57.88	70.11	89.75	104.49
Zn	3344.72	17.60	11.76	37.41	53.71	74.45	91.51	107.59

4.27 Fixed bed column for Chromium

Figures 4.145 to 4.150 illustrate the resulting breakthrough curves for Cr, Cu, Ni and Zn respectively. The heavy metals concentration values at the column exit are plotted as a function of Time or bed volume.

Results on the contact time of Cr on column study at initial metal ion concentration of 10 mg/L by *Chlorella pyrenoidosa, Oscillatoria sp.* and *Spirulina sp.* present in the Figures 4.145 to 4.150. The time required to reach breakthrough for Cr adsorption by *Chlorella pyrenoidosa, Oscillatoria sp.* and *Spirulina sp.* was 120, 150 and 150 minute for 10 mg/L initial metal ion concentration. In the initial concentration of 10 mg/L at the time of 120, 150 and 150 minute percentage of remove of Cr were 77.80 %, 78.60 % and 61.90 % for *Chlorella pyrenoidosa, Oscillatoria sp.* and *Spirulina sp.* respectively.

Percentage of remove of Cr *for the three biomasses at* breakthrough *under study has been established as:*

Chlorella pyrenoidosa > Oscillatoria sp. > Spirulina sp.

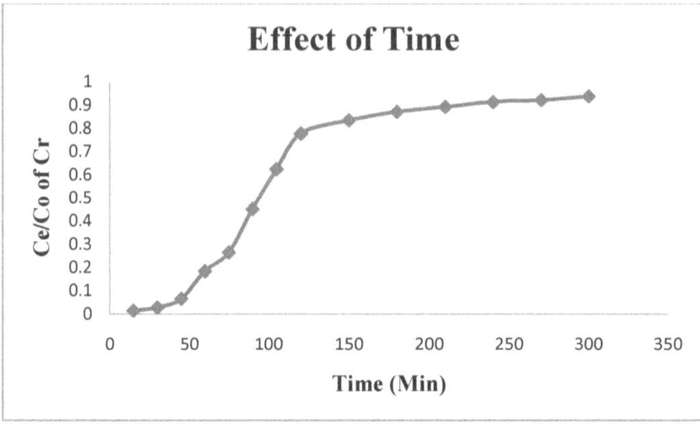

Fig.4.145 Effect of Contact Time on biosorption by *Chlorella pyrenoidosa* (biomass dose=3 g, pH=5, initial ion concentration of Cr, ion concentration = 10 mg/L; temperature = 25 °C; flow rate = 1.67 mL.Min^{-1}, column height= 50 cm).

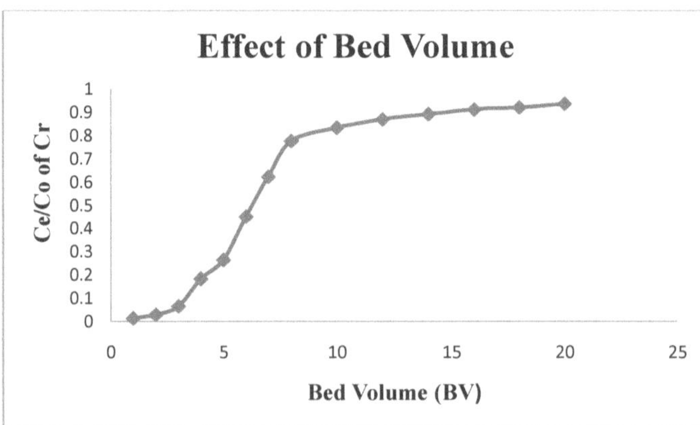

Fig.4.146 Effect of bed volume on biosorption by *Chlorella pyrenoidosa* (biomass dose=3 g, pH=5, initial ion concentration of Cr = 10mg/L; temperature = 25 °C; flow rate = 1.67 mL.Min^{-1}, column height= 50 cm).

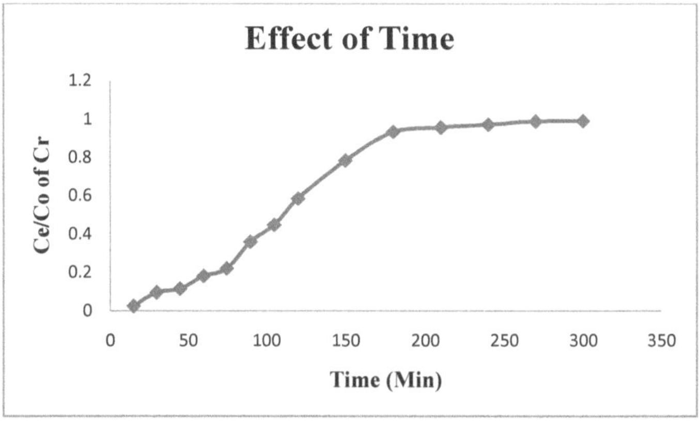

Fig.4.147 Effect of Contact Time on biosorption by *Oscillatoria sp.* (biomass dose=3 g, pH=5, initial ion concentration of Cr, ion concentration = 10 mg/L; temperature = 25 °C; flow rate = 1.67 mL.Min^{-1}, column height= 50 cm).

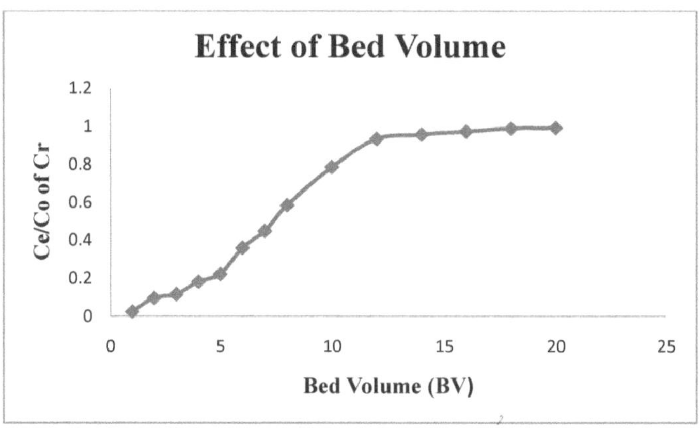

Fig.4.148 Effect of bed volume on biosorption by *Oscillatoria sp.* (biomass dose=3 g, pH=5, initial ion concentration of Cr = 10mg/L; temperature = 25 °C; flow rate = 1.67 mL.Min^{-1}, column height= 50 cm).

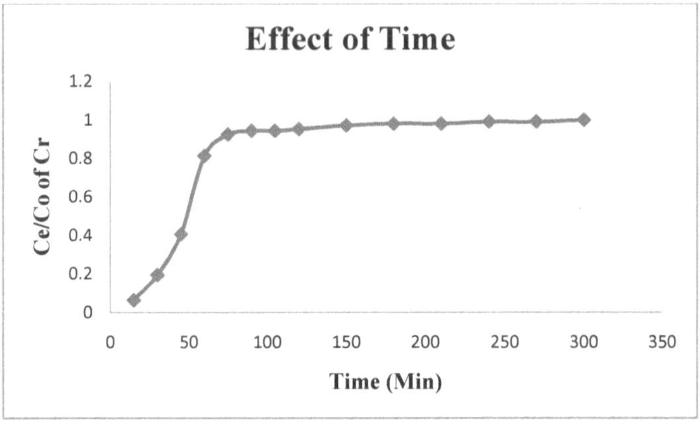

Fig.4.149 Effect of Contact Time on biosorption by *Spirulina sp.* (biomass dose=3 g, pH=5, initial ion concentration of Cr, ion concentration = 10 mg/L; temperature = 25 °C; flow rate = 1.67 mL.Min^{-1}, column height= 50 cm).

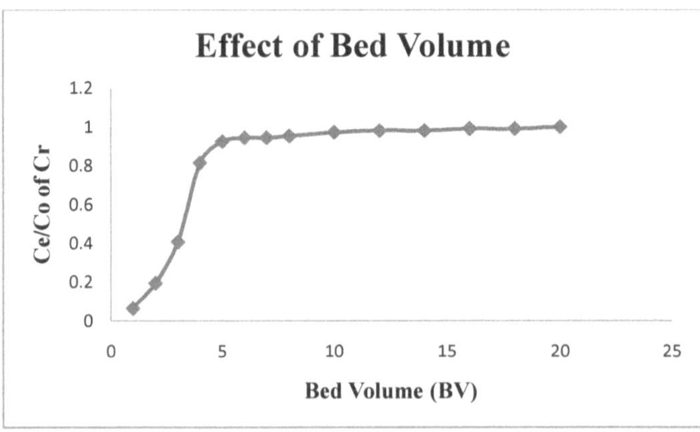

Fig.4.150 Effect of bed volume on biosorption by *Spirulina sp.* (biomass dose=3 g, pH=5, initial ion concentration of Cr = 10mg/L; temperature = 25 °C; flow rate = 1.67 mL.Min^{-1}, column height= 50 cm).

4.28 Fixed bed column for Copper

Results on the contact time of Cu on column study at initial metal ion concentration of 10 mg/L by *Chlorella pyrenoidosa, Oscillatoria sp.* and *Spirulina sp.* present in the Figures 4.151 to 4.156. The time required to reach breakthrough for Cu adsorption by *Chlorella pyrenoidosa, Oscillatoria sp.* and *Spirulina sp.* was 120, 150 and 150 minute for 10 mg/L initial metal ion concentration. In the initial concentration of 10 mg/L at the time of 120, 150 and 150 minute percentage of remove of Cu were 47.00 %, 52.90 % and 62.30 % for *Chlorella pyrenoidosa, Oscillatoria sp.* and *Spirulina sp.* respectively.

Percentage of remove of Cu for *the three biomasses at* breakthrough *under study has been established as:*

Spirulina sp. > *Oscillatoria sp.* > *Chlorella pyrenoidosa*

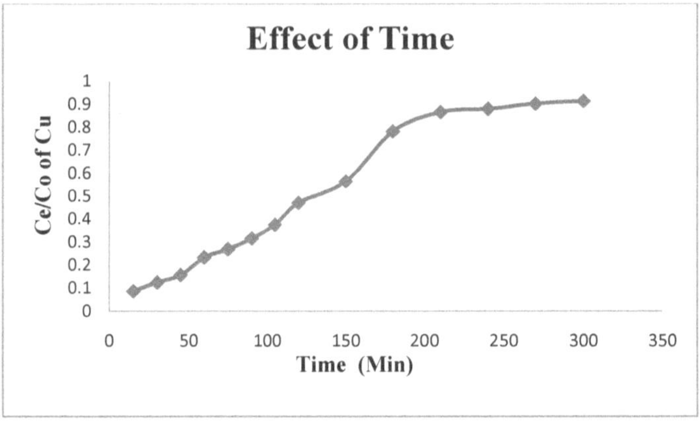

Fig.4.151 Effect of Contact Time on biosorption by *Chlorella pyrenoidosa* (biomass dose=3 g, pH=5, initial ion concentration of Cu, ion concentration = 10 mg/L; temperature = 25 °C; flow rate = 1.67 $mL.Min^{-1}$, column height= 50 cm).

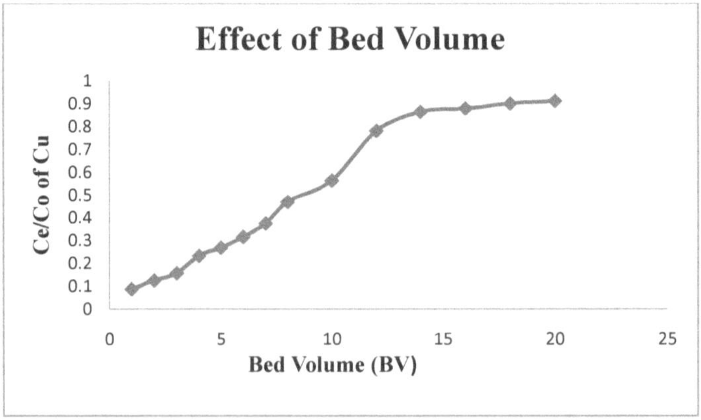

Fig.4.152 Effect of bed volume on biosorption by *Chlorella pyrenoidosa* (biomass dose=3 g, pH=5, initial ion concentration of Cu = 10mg/L; temperature = 25 °C; flow rate = 1.67 $mL.Min^{-1}$, column height= 50 cm).

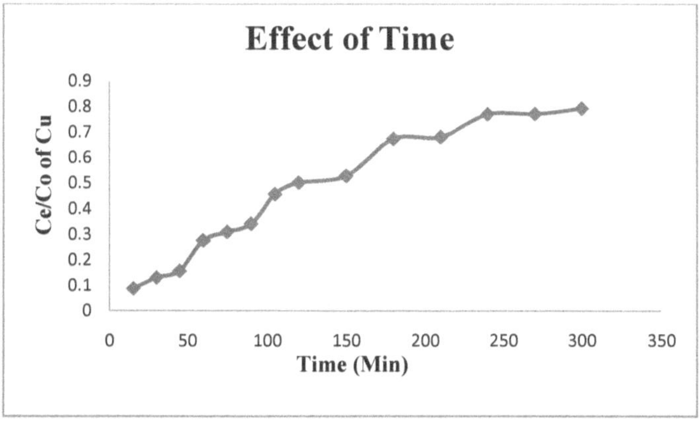

Fig.4.153 Effect of Contact Time on biosorption by *Oscillatoria sp.* (biomass dose=3 g, pH=5, initial ion concentration of Cu, ion concentration = 10 mg/L; temperature = 25 °C; flow rate = 1.67 $mL.Min^{-1}$, column height= 50 cm).

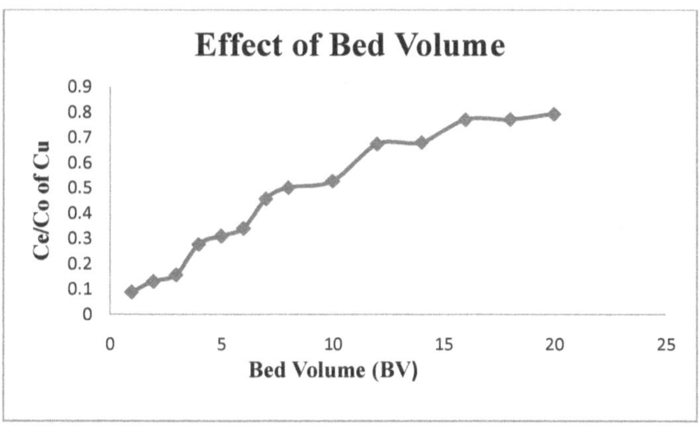

Fig.4.154 Effect of bed volume on biosorption by *Oscillatoria sp.* (biomass dose=3 g, pH=5, initial ion concentration of Cu = 10mg/L; temperature = 25 °C; flow rate = 1.67 $mL.Min^{-1}$, column height= 50 cm).

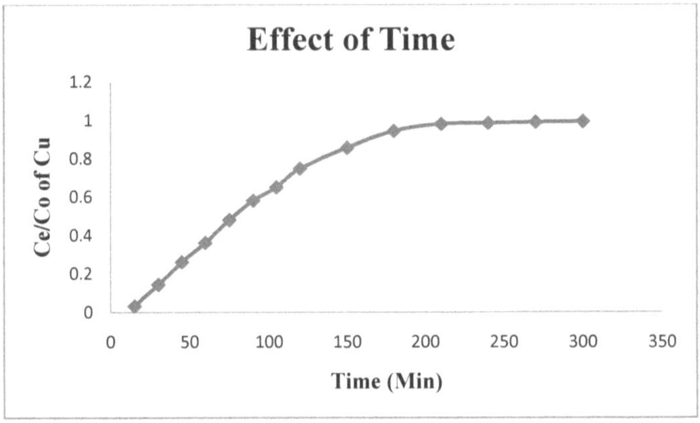

Fig.4.155 Effect of Contact Time on biosorption by *Spirulina sp.* (biomass dose=3 g, pH=5, initial ion concentration of Cu, ion concentration = 10 mg/L; temperature = 25 °C; flow rate = 1.67 mL.Min^{-1}, column height= 50 cm).

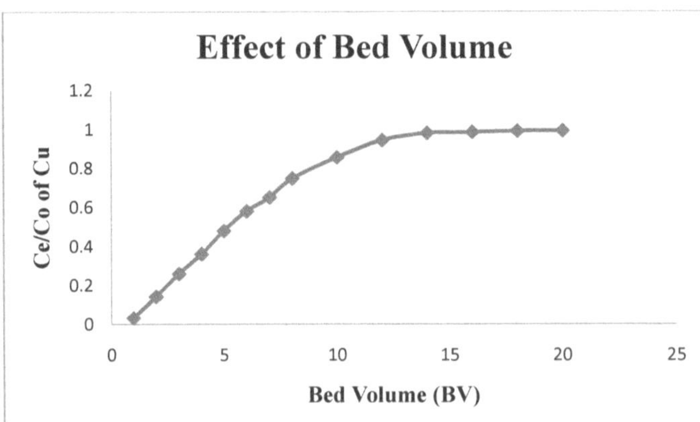

Fig.4.156 Effect of bed volume on biosorption by *Spirulina sp.* (biomass dose=3 g, pH=5, initial ion concentration of Cu = 10mg/L; temperature = 25 °C; flow rate = 1.67 mL.Min^{-1}, column height= 50 cm).

4.29 Fixed bed column for Nickel

Results on the contact time of Ni on column study at initial metal ion concentration of 10 mg/L by *Chlorella pyrenoidosa, Oscillatoria sp.* and *Spirulina sp.* present in the Figures 4.157 to 4.162. The time required to reach breakthrough for Ni adsorption by *Chlorella pyrenoidosa, Oscillatoria sp.* and *Spirulina sp.* was 150,150 and 120 minute for 10 mg/L initial metal ion concentration. In the initial concentration of 10 mg/L at the time of 150, 150 and 120 minute percentage of remove of Ni were 82.30 %, 43.50 % and 60.50 % for *Chlorella pyrenoidosa, Oscillatoria sp.* and *Spirulina sp.* respectively.

Percentage of remove of Ni for *the three biomasses at* breakthrough *under study has been established as:*

Chlorella pyrenoidosa > Spirulina sp. > Oscillatoria sp.

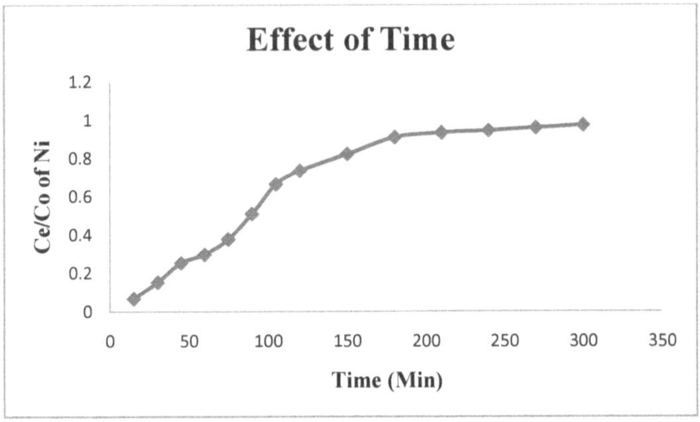

Fig.4.157 Effect of Contact Time on biosorption by *Chlorella pyrenoidosa* (biomass dose=3 g, pH=5, initial ion concentration of Ni, ion concentration = 10 mg/L; temperature = 25 °C; flow rate = 1.67 mL.Min^{-1}, column height= 50 cm).

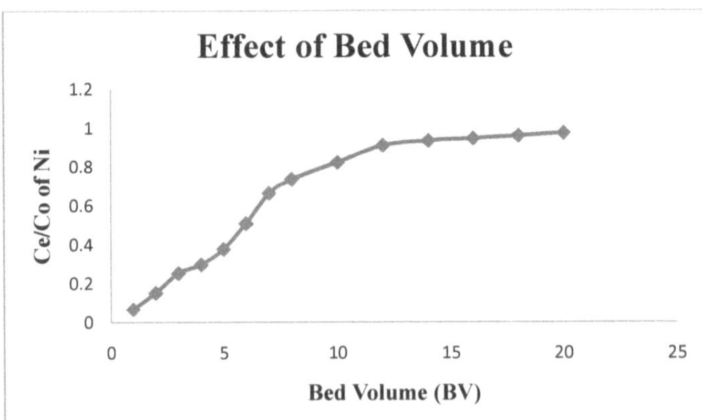

Fig.4.158 Effect of bed volume on biosorption by *Chlorella pyrenoidosa* (biomass dose=3 g, pH=5, initial ion concentration of Ni = 10mg/L; temperature = 25 °C; flow rate = 1.67 mL.Min^{-1}, column height= 50 cm).

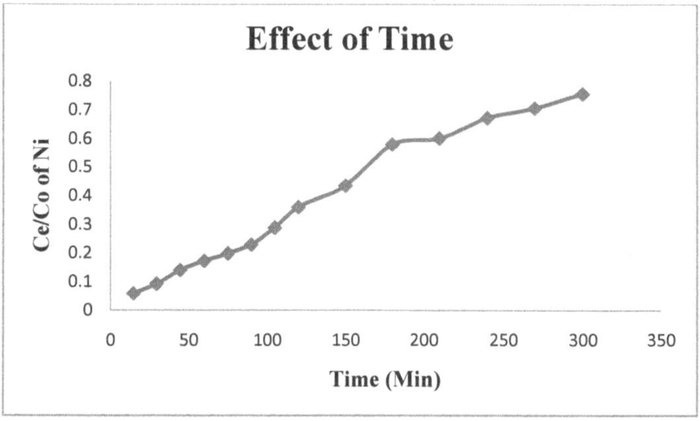

Fig.4.159 Effect of Contact Time on biosorption by *Oscillatoria sp.* (biomass dose=3 g, pH=5, initial ion concentration of Ni, ion concentration = 10 mg/L; temperature = 25 ˚C; flow rate = 1.67 mL.Min^{-1}, column height= 50 cm).

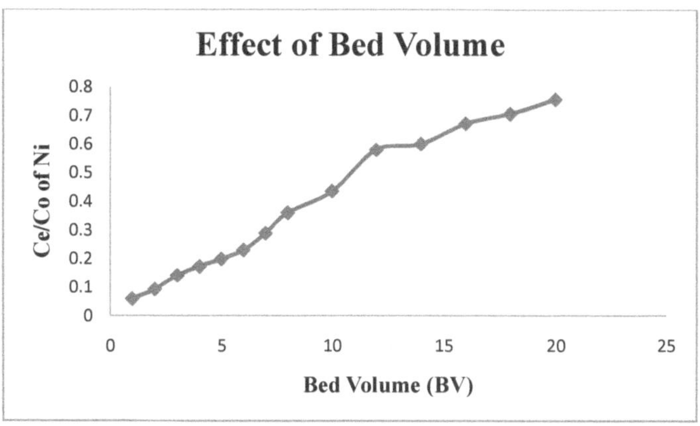

Fig.4.160 Effect of bed volume on biosorption by *Oscillatoria sp.* (biomass dose=3 g, pH=5, initial ion concentration of Ni = 10mg/L; temperature = 25 ˚C; flow rate = 1.67 mL.Min^{-1}, column height= 50 cm).

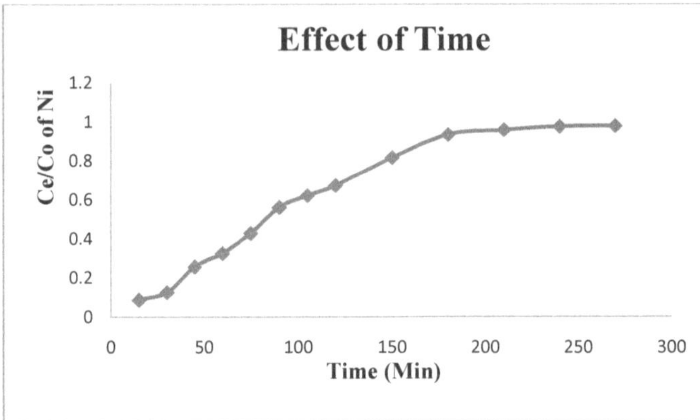

Fig.4.161 Effect of Contact Time on biosorption by *Spirulina sp.* (biomass dose=3 g, pH=5, initial ion concentration of Ni, ion concentration = 10 mg/L; temperature = 25 °C; flow rate = 1.67 $mL.Min^{-1}$, column height= 50 cm).

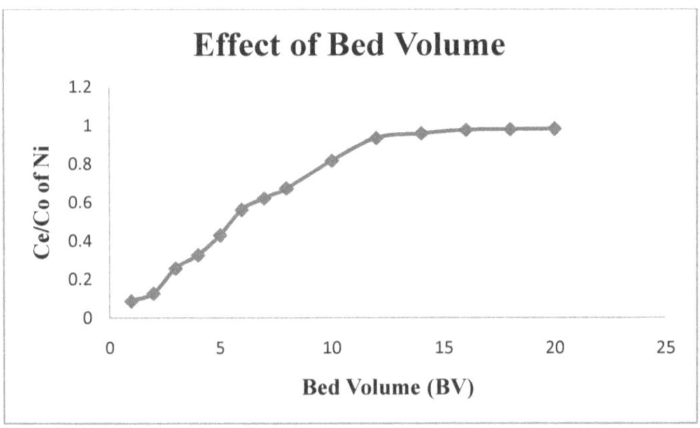

Fig.4.162 Effect of bed volume on biosorption by *Spirulina sp.* (biomass dose=3 g, pH=5, initial ion concentration of Ni = 10mg/L; temperature = 25 °C; flow rate = 1.67 $mL.Min^{-1}$, column height= 50 cm).

4.30 Fixed bed column for Zinc

Results on the contact time of Zn on column study at initial metal ion concentration of 10 mg/L by *Chlorella pyrenoidosa, Oscillatoria sp.* and *Spirulina sp.* present in the Figures 4.163 to 4.168. The time required to reach breakthrough for Zn adsorption by *Chlorella pyrenoidosa, Oscillatoria sp.* and *Spirulina sp.* was 120,120 and 150 minute for 10 mg/L initial metal ion concentration. In the initial concentration of 10 mg/L at the time of 120, 120 and 150 minute percentage of remove of Zn were 54.90 %, 58.80 % and 64.50 % for *Chlorella pyrenoidosa, Oscillatoria sp.* and *Spirulina sp.* respectively.

Percentage of remove of Zn for *the three biomasses at* breakthrough *under study has been established as:*

Spirulina sp. > *Oscillatoria sp.* > *Chlorella pyrenoidosa*

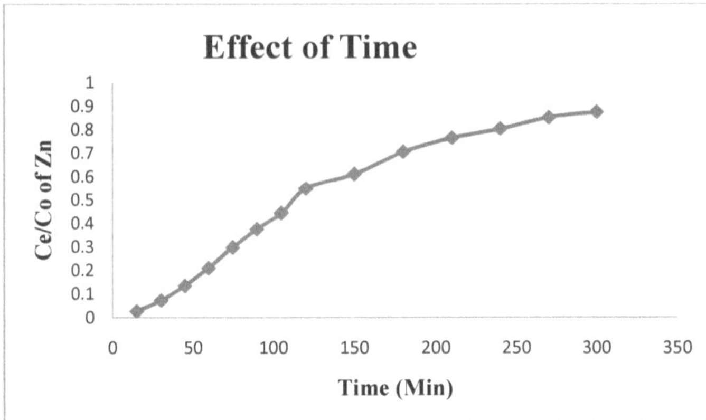

Fig.4.163 Effect of Contact Time on biosorption by *Chlorella pyrenoidosa* (biomass dose=3 g, pH=5, initial ion concentration of Zn, ion concentration = 10 mg/L; temperature = 25 °C; flow rate = 1.67 $mL.Min^{-1}$, column height= 50 cm).

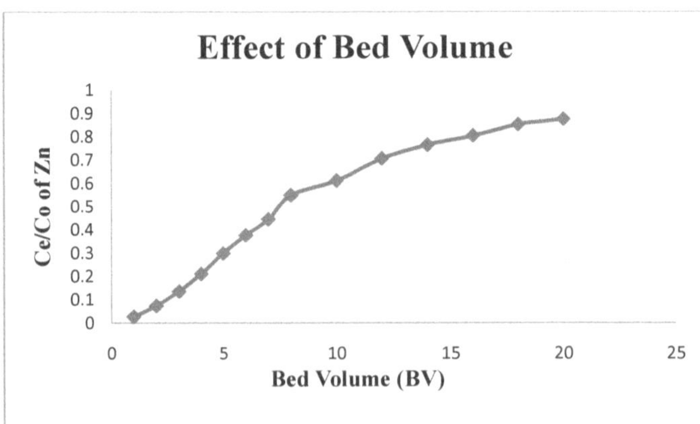

Fig.4.164 Effect of bed volume on biosorption by *Chlorella pyrenoidosa* (biomass dose=3 g, pH=5, initial ion concentration of Zn = 10mg/L; temperature = 25 °C; flow rate = 1.67 $mL.Min^{-1}$, column height= 50 cm).

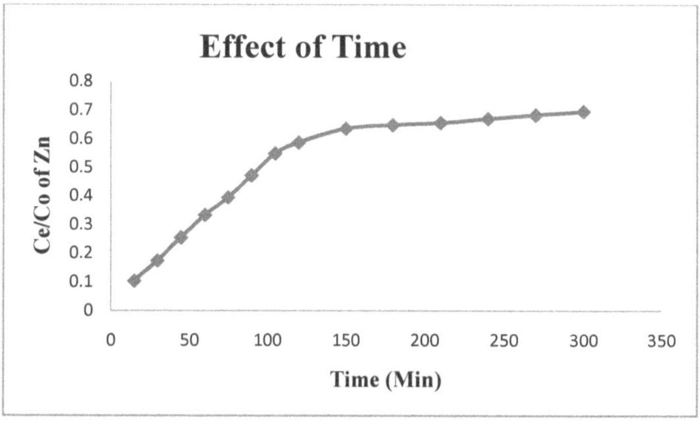

Fig.4.165 Effect of Contact Time on biosorption by *Oscillatoria sp.* (biomass dose=3 g, pH=5, initial ion concentration of Zn, ion concentration = 10 mg/L; temperature = 25 ˚C; flow rate = 1.67 $mL.Min^{-1}$, column height= 50 cm).

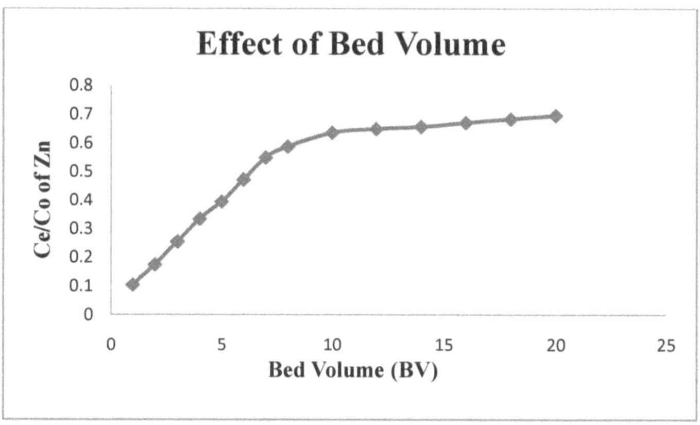

Fig.4.166 Effect of bed volume on biosorption by *Oscillatoria sp.* (biomass dose=3 g, pH=5, initial ion concentration of Zn = 10mg/L; temperature = 25 ˚C; flow rate = 1.67 $mL.Min^{-1}$, column height= 50 cm).

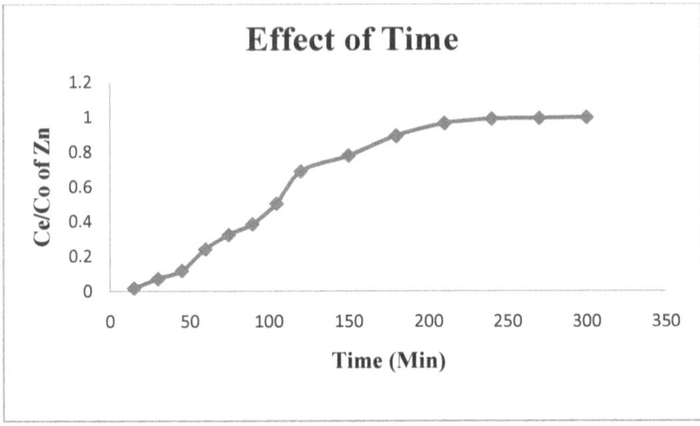

Fig.4.167 Effect of Contact Time on biosorption by *Spirulina sp.* (biomass dose=3 g, pH=5, initial ion concentration of Zn, ion concentration = 10 mg/L; temperature = 25 °C; flow rate = 1.67 mL.Min^{-1}, column height= 50 cm).

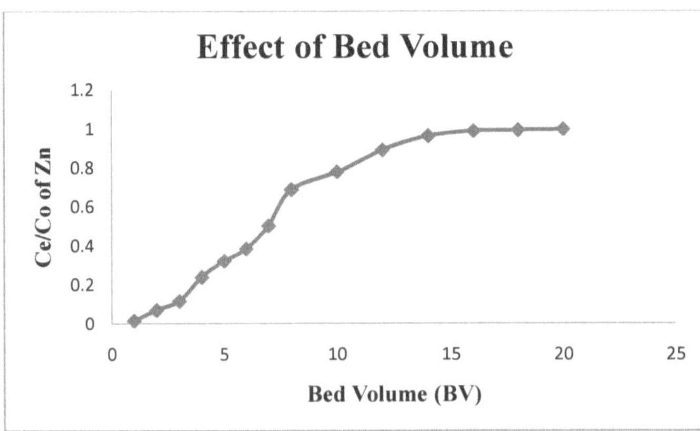

Fig.4.168 Effect of bed volume on biosorption by *Spirulina sp.* (biomass dose=3 g, pH=5, initial ion concentration of Zn = 10mg/L; temperature = 25 °C; flow rate = 1.67 mL.Min^{-1}, column height= 50 cm).

4.31 Effect of time on column study

Results on the contact time of heavy metals on column study at initial metal ion concentration of 10 mg/L by *Chlorella pyrenoidosa, Oscillatoria sp.* and *Spirulina sp.* present in the Figures 4.169 to 4.174.

In effect of time and initial concentration of 10 mg / L for Cr, Cu, Ni and Zn, percentage of remove of heavy metals and bed volume *for the Chlorella pyrenoidosa under study has been established as:*

Ni > Cr > Cu > Zn

In effect of time and initial concentration of 10 mg / L for Cr, Cu, Ni and Zn, percentage of remove of heavy metals and bed volume *for the Oscillatoria sp. under study has been established as:*

Cr > Cu > Ni > Zn

In effect of time and initial concentration of 10 mg / L for Cr, Cu, Ni and Zn, percentage of remove of heavy metals and bed volume *for the Spirulina sp. under study has been established as:*

Zn > Cu > Ni > Cr

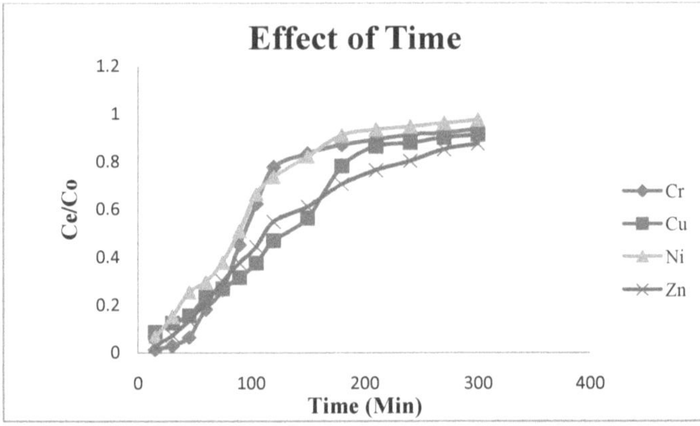

Fig.4.169 Effect of Contact Time on biosorption by *Chlorella pyrenoidosa* (biomass dose=3 g, pH=5, initial Cr, Cu, Ni and Zn ion concentration = 10 mg/L; temperature = 25 °C; flow rate = 1.67 mL.Min^{-1}, column height= 50 cm).

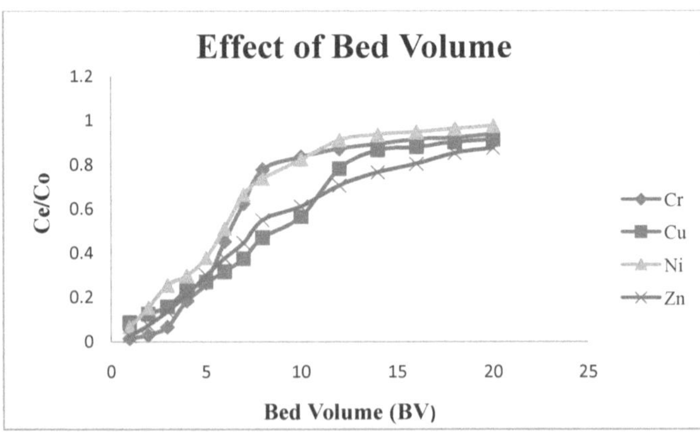

Fig.4.170 Effect of bed volume on biosorption by Chlorella (biomass dose=3 g, pH=5, initial Cr, Cu, Ni and Zn ion concentration = 10mg/L; temperature = 25 °C; flow rate = 1.67 mL.Min^{-1}, column height= 50 cm).

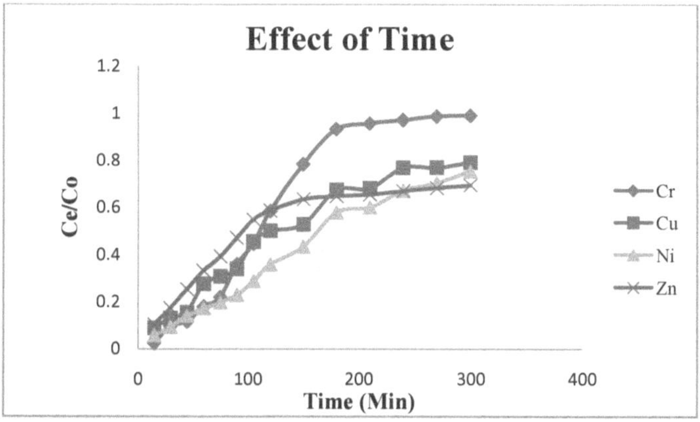

Fig.4.171 Effect of Contact Time on biosorption by *Oscillatoria sp.* (biomass dose=3 g, pH=5, initial Cr, Cu, Ni and Zn ion concentration = 10 mg/L; temperature = 25 ˚C; flow rate = 1.67 mL.Min^{-1}, column height= 50 cm).

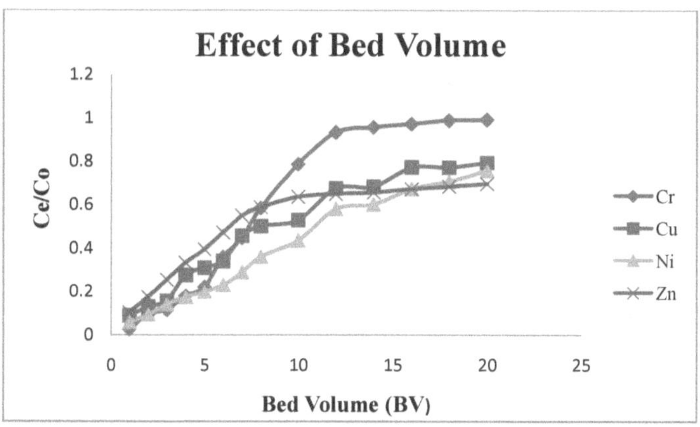

Fig.4.172 Effect of bed volume on biosorption by *Oscillatoria sp.* (biomass dose=3 g, pH=5, initial Cr, Cu, Ni and Zn ion concentration = 10mg/L; temperature = 25 ˚C; flow rate = 1.67 mL.Min^{-1}, column height= 50 cm).

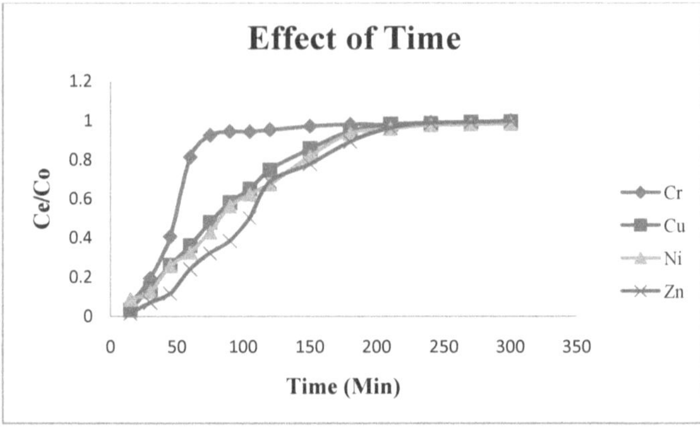

Fig.4.173 Effect of Contact Time on biosorption by *Spirulina sp.* (biomass dose=3 g, pH=5, initial Cr, Cu, Ni and Zn ion concentration = 10 mg/L; temperature = 25 ˚C; flow rate = 1.67 mL.Min^{-1}, column height= 50 cm).

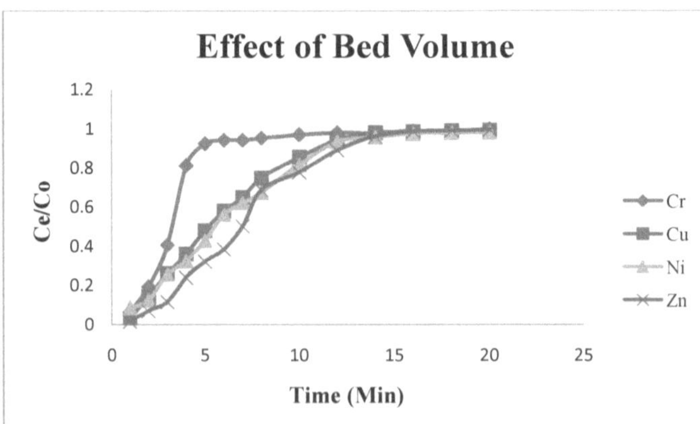

Fig.4.174 Effect of bed volume on biosorption by *Spirulina sp.* (biomass dose=3 g, pH=5, initial Cr, Cu, Ni and Zn ion concentration = 10mg/L; temperature = 25 ˚C; flow rate = 1.67 mL.Min^{-1}, column height= 50 cm).

4.32 Desorption studies

Results of the desorption of heavy metals by using of 0.1 M EDTA, HCL, HNO_3 summarized on the Table 4.21. All of the result showed that the nitric acid was the best chemical solution for desorption process.

$$HNO_3 > HCL > EDTA$$

Table 4.21 The desorption efficiency of different desorbent

Desorbent	EDTA (0.1M)	HCL(0.1M)	HNO_3(0.1M)
% Desorption of Cr	72.23 ± 3.15	89.57 ±3.39	95.04 ±3.10
% Desorption of Cu	64.88 ±4.66	90.08 ±4.15	92.69 ±2.61
% Desorption of Ni	61.48 ± 2.98	93.17 ±3.55	97.70 ±3.25
% Desorption of Zn	76 .14±4.12	91.36 ±4.43	94.84 ±4.01

4.33 FTIR Spectroscopy

4.33.1 FTIR Spectroscopy of *Chlorella pyrenoidosa*

Spectrums of *Chlorella pyrenoidosa* present in the Figures 4.175 to 4.183 that revealed the presence of hydroxyl, amino, carboxylic and carbonyl groups. The presence of OH group along with carbonyl group confirmed the presence of carboxylic acid groups in the biosorbent. The presence of NH group and OH group along with carbonyl group might be attributed the presence of amino acid groups in the biosorbent. Assignment of bands to functional group on the surface of *Chlorella pyrenoidosa* before and after biosorption of chromium, copper, nickel and zinc as observed from FTIR spectroscopy summarized on the **Table4.22**.

Table4.22 Assignment of bands to functional group on the surface of *Chlorella pyrenoidosa* as observed from FTIR spectroscopy.

Wave Numbers (Cm^{-1})					Assignment
Befor	**Chromium**	**Copper**	**Nickel**	**Zinc**	
3433.41 3325.39 3255.95 3213.51 3174.94	3450.77 3317.67 3174.94	3371.68 3173.01	3385.18 3173.01 3078.49	3373.61 3174.94 3091.99	O-H Stretching / N-H Stretching
2924.18	2924.18	2924.18	2926.11	2924.18	C-H Stretching Aliphatic
2854.74 2746.73	2854.74	2856.67 2748.65	2854.74 2742.87	2854.74 2744.80	C-H Stretching
1737.92	1741.78	1735.99	1732.13	1732.13	C=O Stretching vibration
1639.55	1647.26 1618.33	1670.41 1633.76	1645.33	1645.33 1618.33	C=O Stretching vibration / N-H Stretching vibration
			1518.03		
1450.52	1452.45	1460.16	1452.45	1454.38	CH_2Bonding vibration
1317.43	1365.65		1315.50	1315.50	CH_3 Bonding vibration
1033.88	1033.88	1035.81 1012.66	1035.81	1031.95	C-O Stretching
756.12	758.05	754.19	758.05		C-X (X = F, Cl, Br and I)

4.33.2 FTIR Spectroscopy of *Oscillatoria sp.*

Spectrums of *Oscillatoria sp.* present in the Figures 4.184 to 4.192 that revealed the presence of hydroxyl, amino, carboxylic and carbonyl groups. The presence of OH group along with carbonyl group confirmed the presence of carboxylic acid groups in the biosorbent. The presence of NH group and OH group along with carbonyl group might be attributed the presence of amino acid groups in the biosorbent. The stretching vibration of OH, NH and C=O groups shifted for a certain extent in addition, which indicated that these three groups possibly involved in the biosorption. Assignment of bands to functional group on the surface of *Oscillatoria sp.* before and after biosorption of chromium, copper, nickel and zinc as observed from FTIR spectroscopy summarized on the **Table4.23**.

Table4.23 Assignment of bands to functional group on the surface of *Oscillatoria sp.* as observed from FTIR spectroscopy.

Wave Numbers (Cm^{-1})					
Befor	**Chromium**	**Copper**	**Nickel**	**Zinc**	**Assignment**
3309.96	3576.14 3475.84 3404.47 3362.04 3144.07	3493.20 3261.74	3543.35 3452.70 3250.16	3489.34 3279.10 3082.35	O-H Stretching / N-H Stretching
2929.97	2933.83	2924.18	2906.82	2933.83	C-H Stretching Aliphatic
2810.38	2858.60	2860.53	2854.74	2860.53 2733.22	C-H Stretching
1660.77	1670.41 1633.76	1635.69	1627.97	1668.48 1624.12	C=O Stretching vibration / N-H Stretching vibration
1514.17	1519.96	1543.10	1570.11	1527.67	
1450.52	1444.73	1442.80	1460.16	1404.22	CH_2Bonding vibration
1375.29	1377.22	1381.08	1126.47	1388.79	CH_3 Bonding vibration
1039.67	1078.24 1047.38	1049.31	1031.95 1003.95	1043.52	C-O Stretching
765.77		707.90	744.55	821.70	C-X (X = F, Cl, Br and I)

4.33.3 FTIR Spectroscopy of *Spirulina sp.*

Spectrums of *Spirulina sp.* present in the Figures 4.193 to 4.201 that revealed the presence of hydroxyl, amino, carboxylic and carbonyl groups. The presence of OH group along with carbonyl group confirmed the presence of carboxylic acid groups in the biosorbent. The presence of NH group and OH group along with carbonyl group might be attributed the presence of amino acid groups in the biosorbent. The stretching vibration of OH, NH and C=O groups shifted for a certain extent in addition, which indicated that these three groups possibly involved in the biosorption. Assignment of bands to functional group on the surface of *Spirulina sp.* before and after biosorption of chromium, copper, nickel and zinc as observed from FTIR spectroscopy summarized on the **Table4.24**.

Table4.24 Assignment of bands to functional group on the surface of *Spirulina sp.* as observed from FTIR spectroscopy.

Wave Numbers (Cm^{-1})					
Befor	**Chromium**	**Copper**	**Nickel**	**Zinc**	**Assignment**
3350.46	3308.03	3292.60	3315.74	3290.67	
3217.37	3174.94	3213.51	3255.95		O-H Stretching /
			3213.51	3174.94	N-H Stretching
3090.07		3082.35			C-H Stretching Aromatic
2956.97	2955.04	2926.11	2956.97	2956.97	C-H Stretching Aliphatic
2926.11	2926.11		2926.11	2926.11	
2856.67	2854.74	2854.74	2854.74	2854.74	C-H Stretching
	1730.21				C=O Stretching vibration
1672.34	1647.26	1647.26	1645.33	1633.76	C=O Stretching vibration /
1627.97	1622.19				N-H Stretching vibration
1531.53	1539.25	1537.32	1533.46	1539.25	
				1508.38	
1446.66	1454.38	1454.38	1448.59	1456.30	CH_2Bonding vibration
				1375.29	CH_3 Bonding vibration
1039.67	1043.52	1012.66	1037.74	1043.52	C-O Stretching
777.34	773.48	700.18		702.11	C-X (X = F, Cl, Br and I)

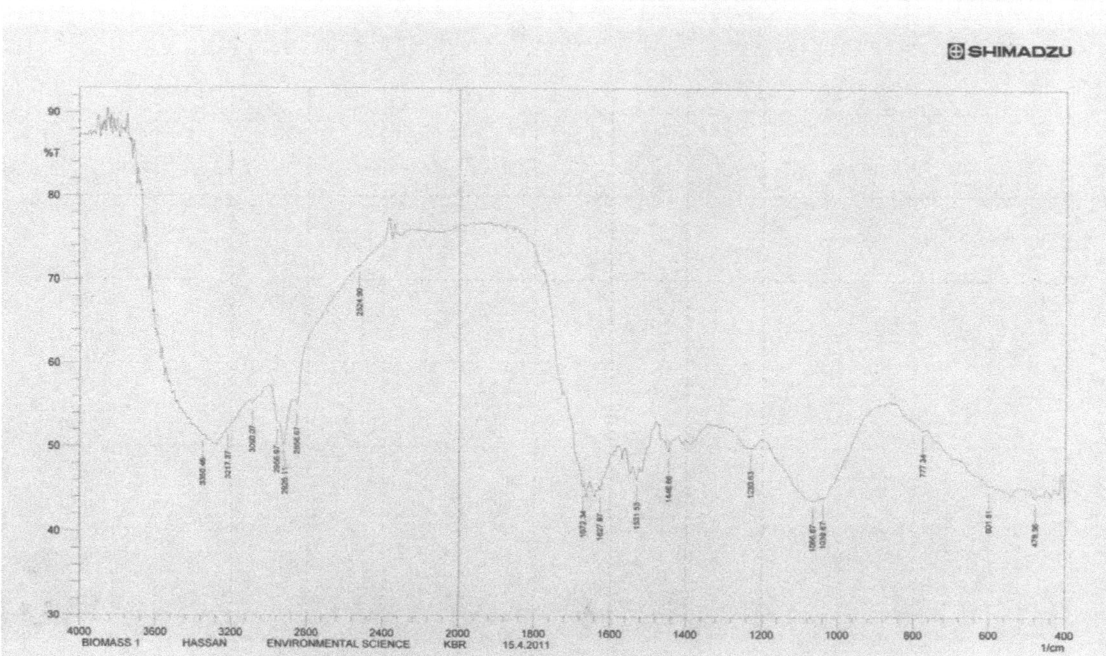

Fig.4.175 FTIR Spectrum of the *Chlorella pyrenoidosa* (before biosorption).

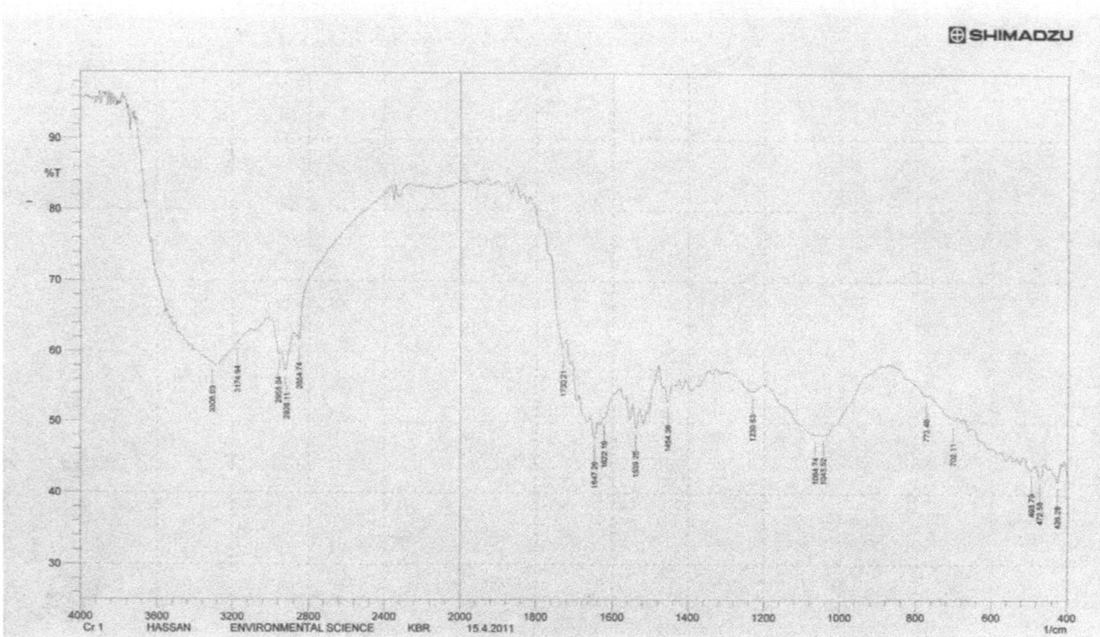

Fig.4.176 FTIR Spectrum of the *Chlorella pyrenoidosa* (after biosorption of Cr).

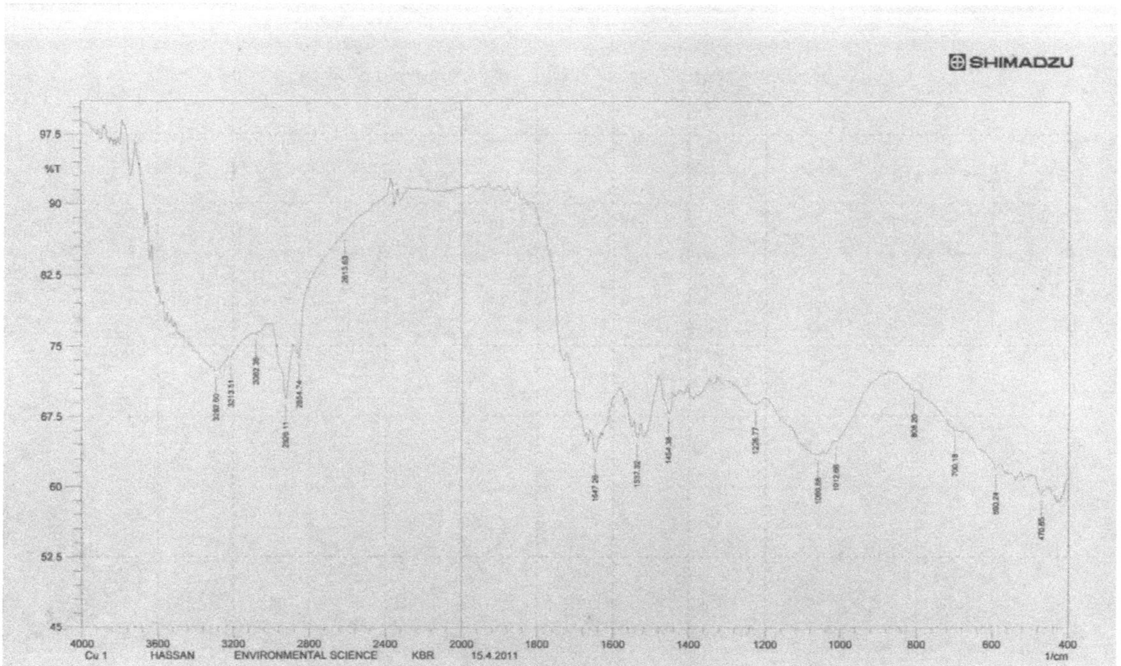

Fig.4.177 FTIR Spectrum of the *Chlorella pyrenoidosa* (after biosorption of Cu).

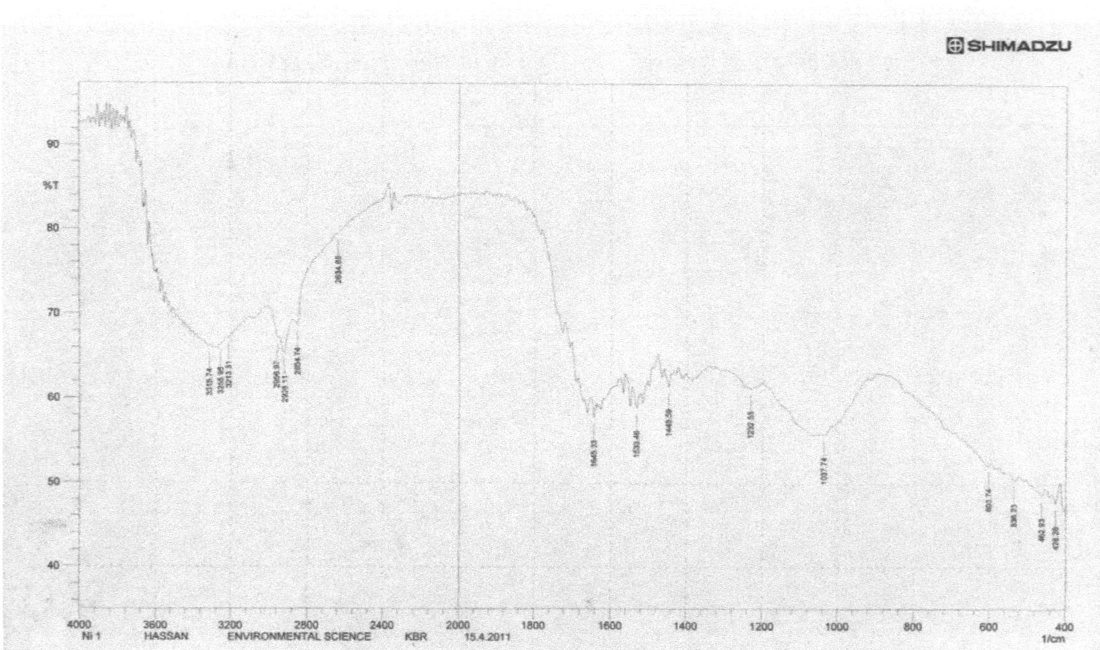

Fig.4.178 FTIR Spectrum of the *Chlorella pyrenoidosa* (after biosorption of Ni).

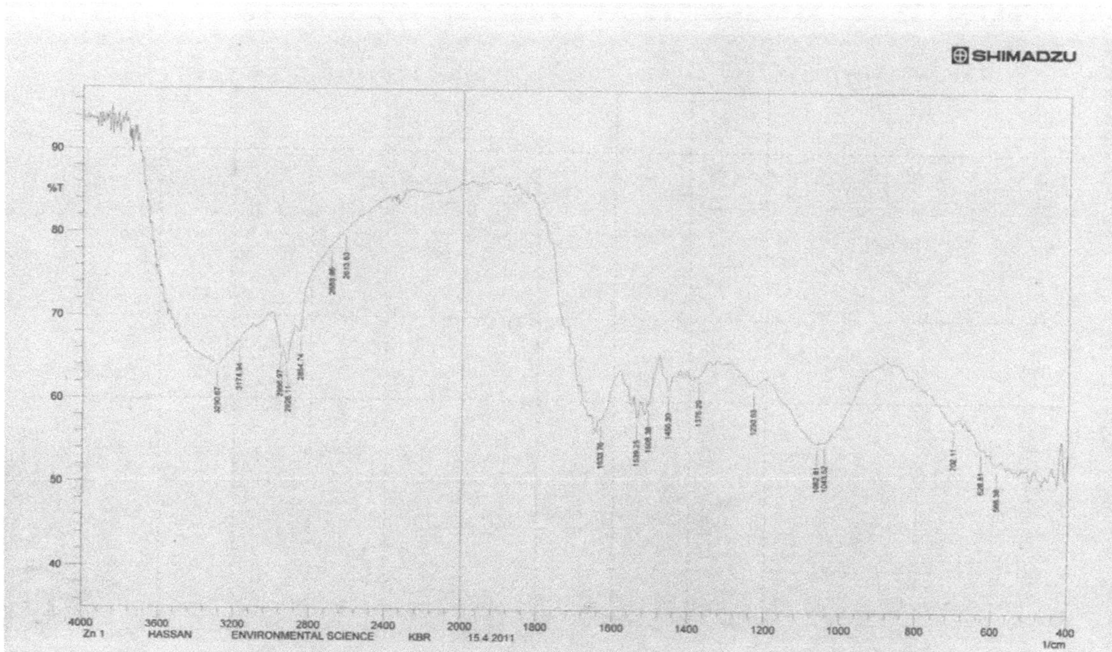

Fig.4.179 FTIR Spectrum of the *Chlorella pyrenoidosa* (after biosorption of Zn).

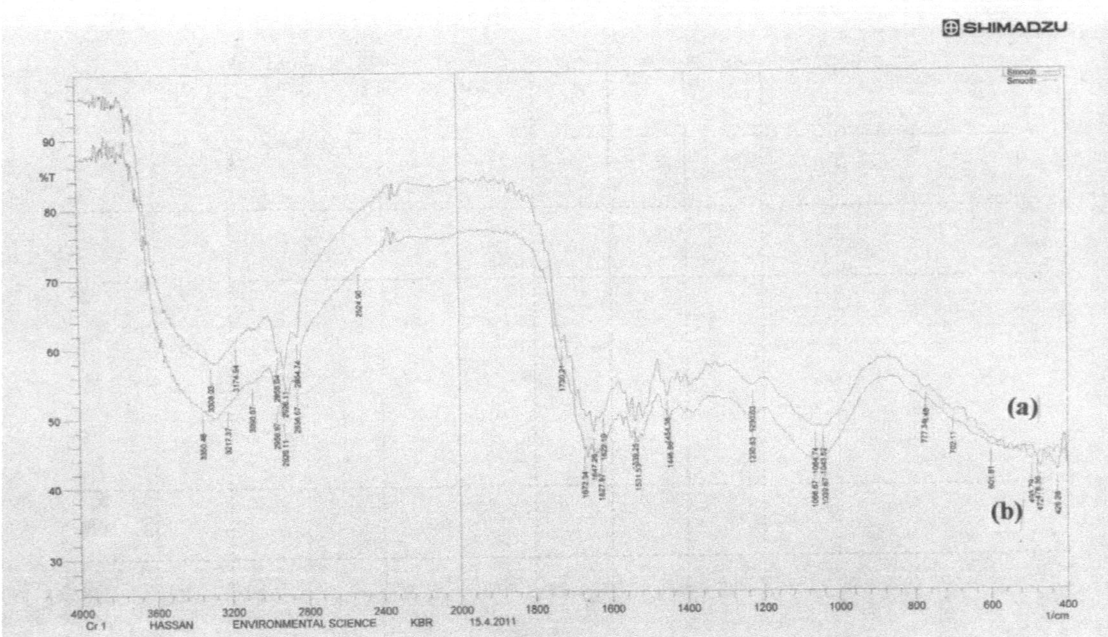

Fig.4.180 FTIR Spectrum of the *Chlorella pyrenoidosa* (before **(b)** and after **(a)** biosorption of Cr).

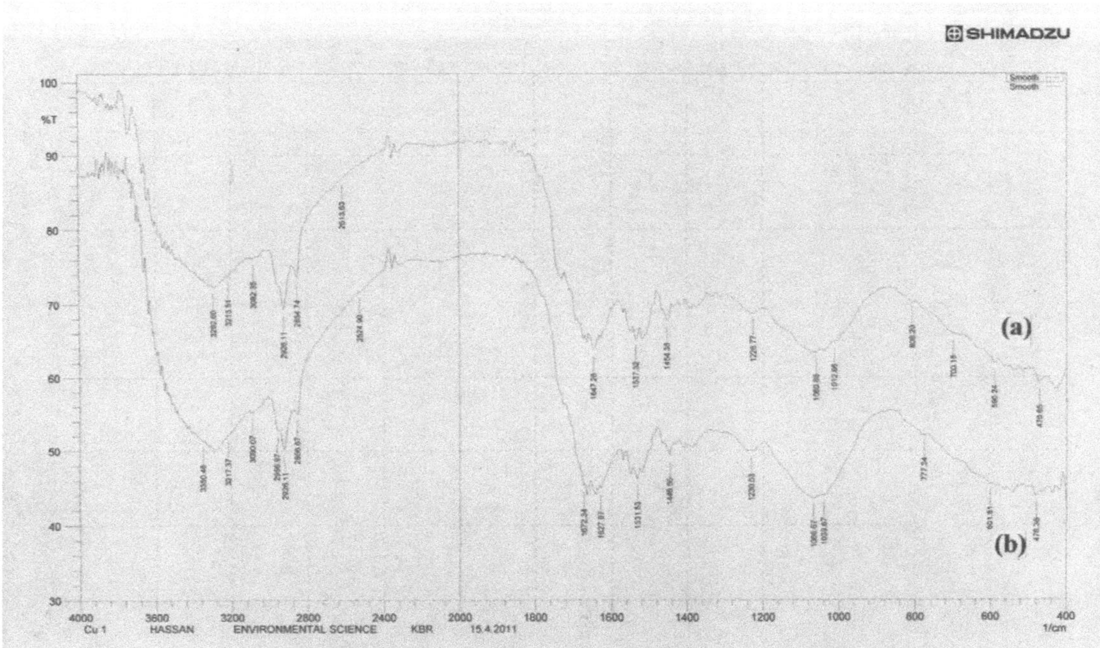

Fig.4.181 FTIR Spectrum of the *Chlorella pyrenoidosa* (before **(b)** and after **(a)** biosorption of Cu).

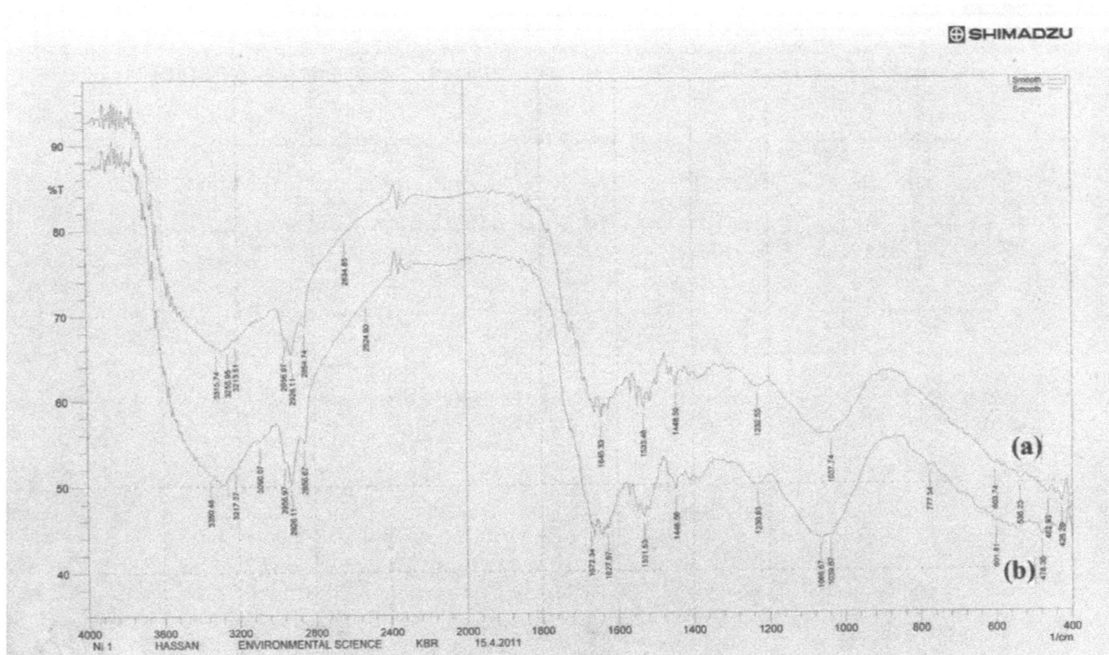

Fig.4.182 FTIR Spectrum of the *Chlorella pyrenoidosa* (before **(b)** and after **(a)** biosorption of Ni).

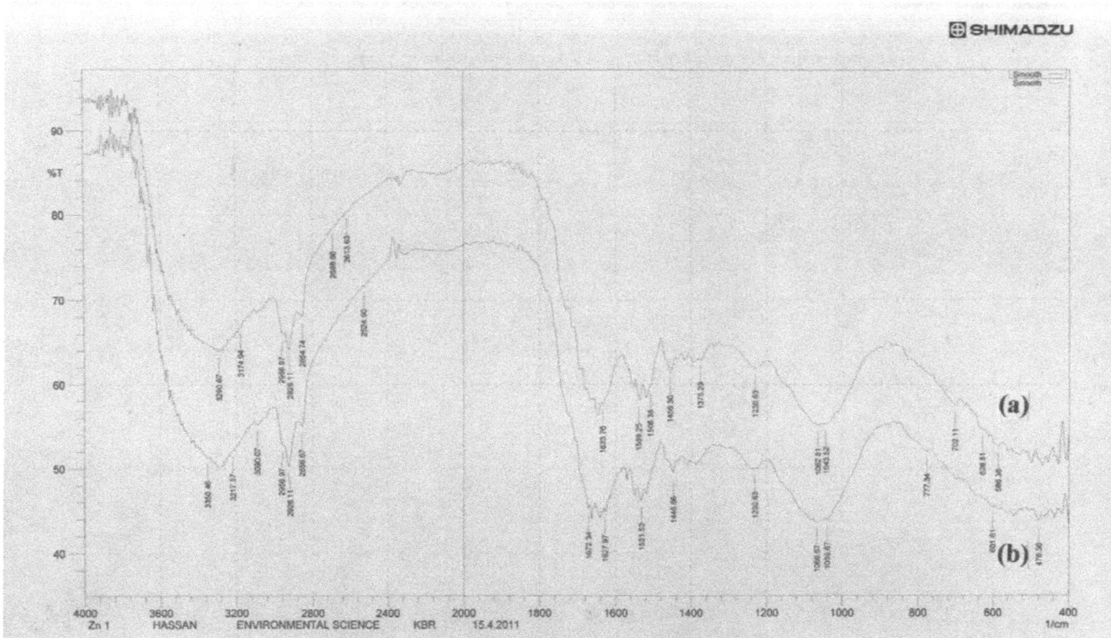

Fig.4.183 FTIR Spectrum of the *Chlorella pyrenoidosa* (before **(b)** and after **(a)** biosorption of Zn).

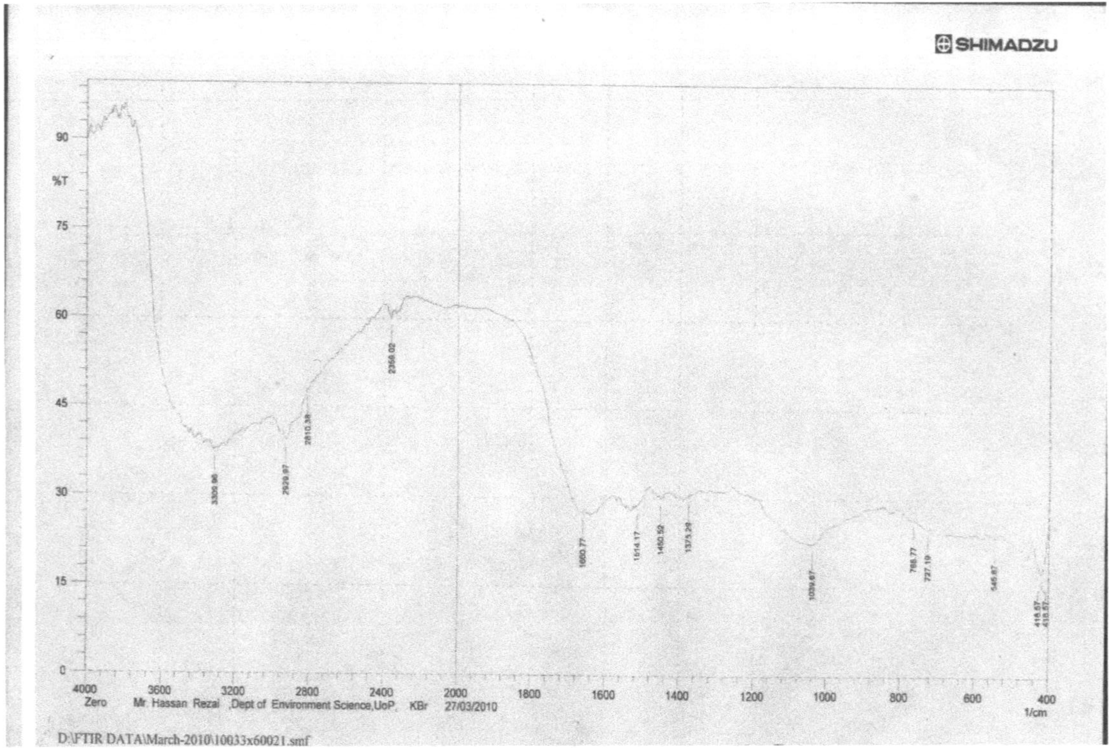

Fig.4.184 FTIR Spectrum of the *Oscillatoria sp.* (before biosorption).

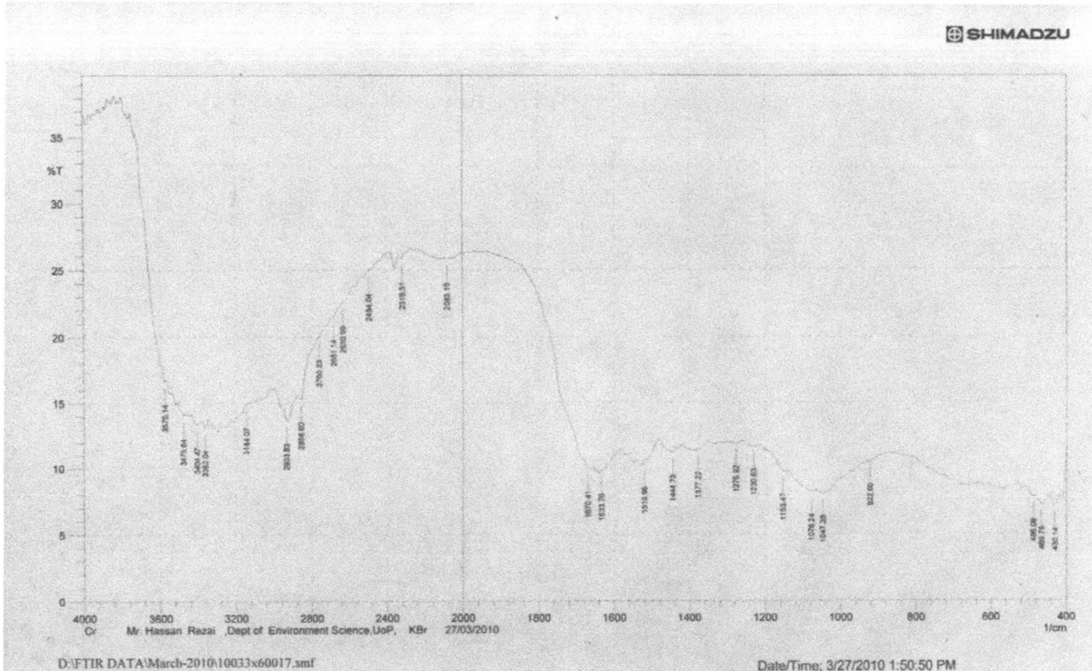

Fig.4.185 FTIR Spectrum of the *Oscillatoria sp.* (after biosorption of Cr).

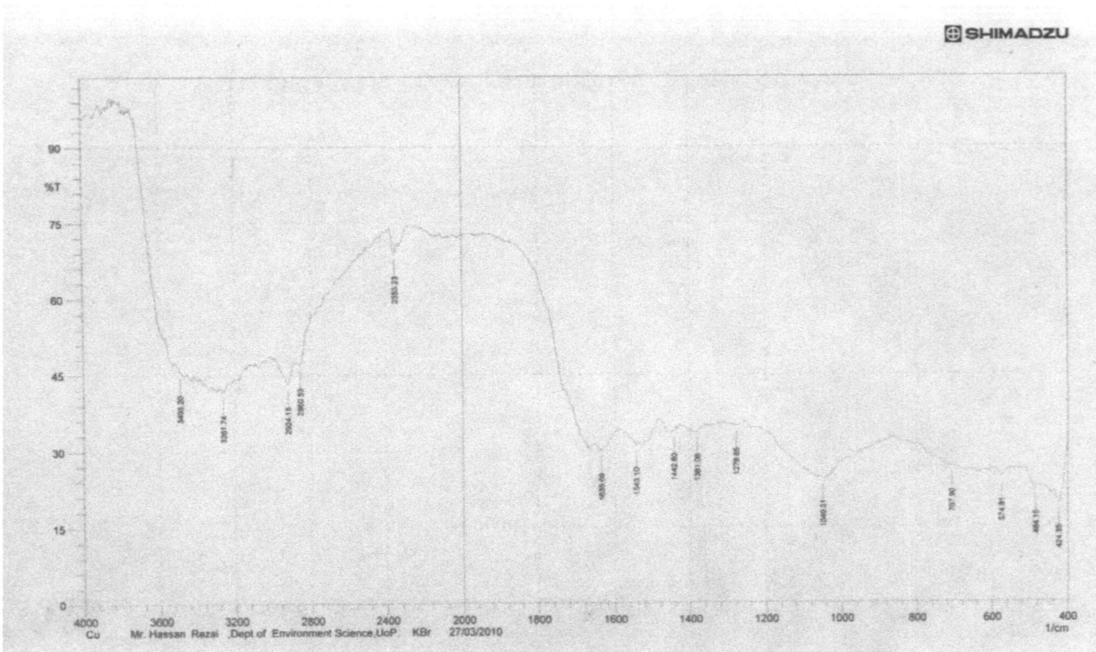

Fig.4.186 FTIR Spectrum of the *Oscillatoria sp.* (after biosorption of Cu).

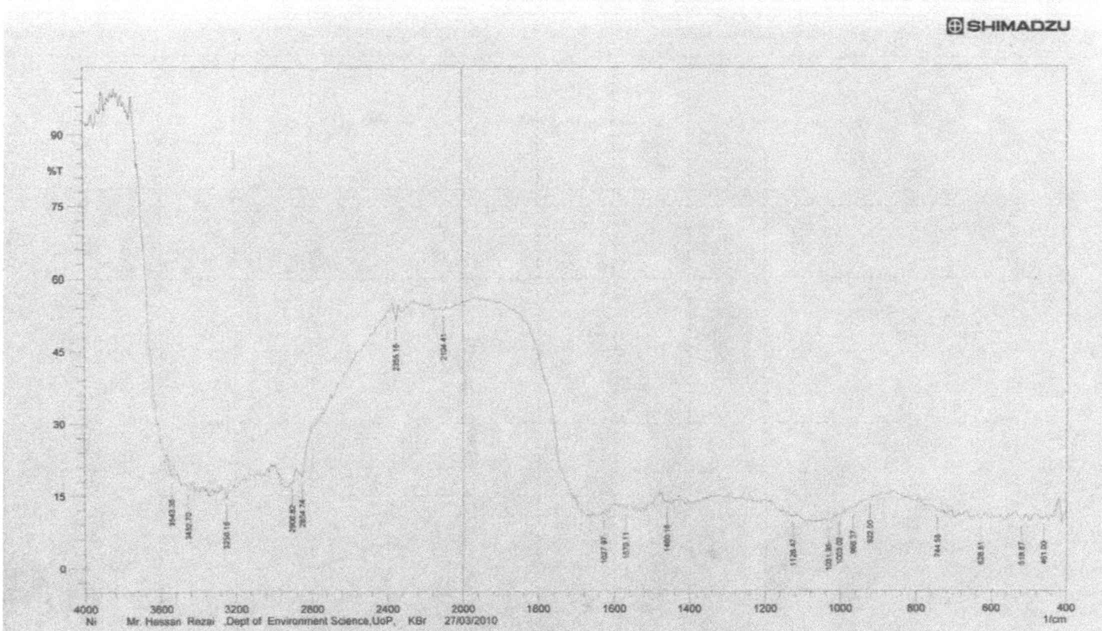

Fig.4.187 FTIR Spectrum of the *Oscillatoria sp.* (after biosorption of Ni).

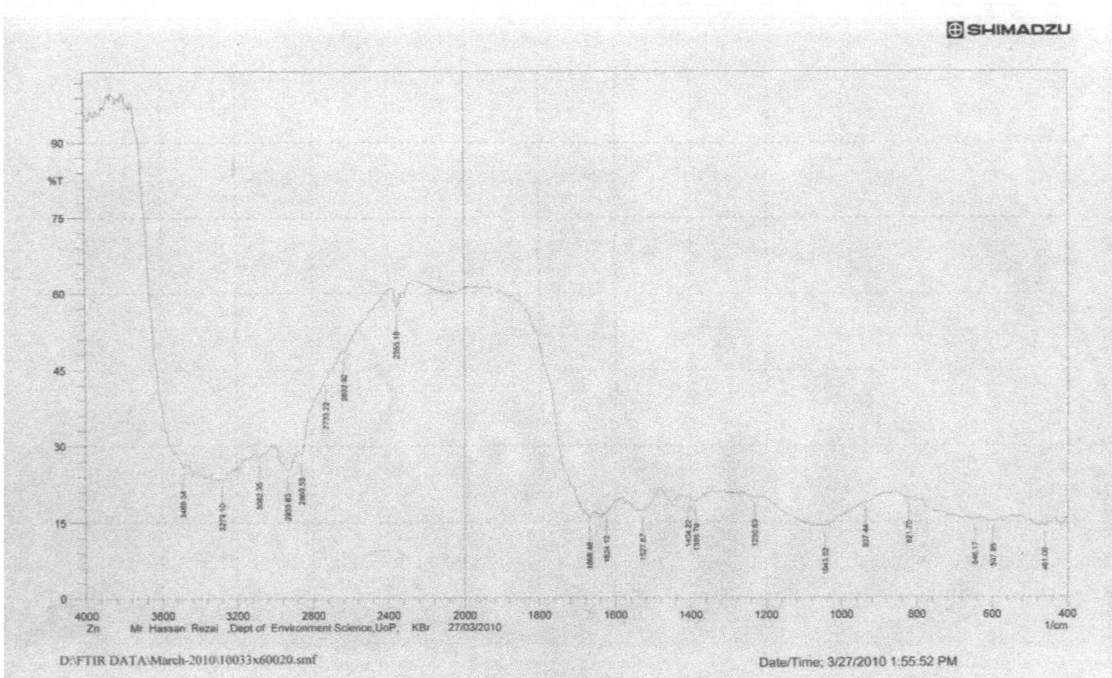

Fig.4.188 FTIR Spectrum of the *Oscillatoria sp.* (after biosorption of Zn).

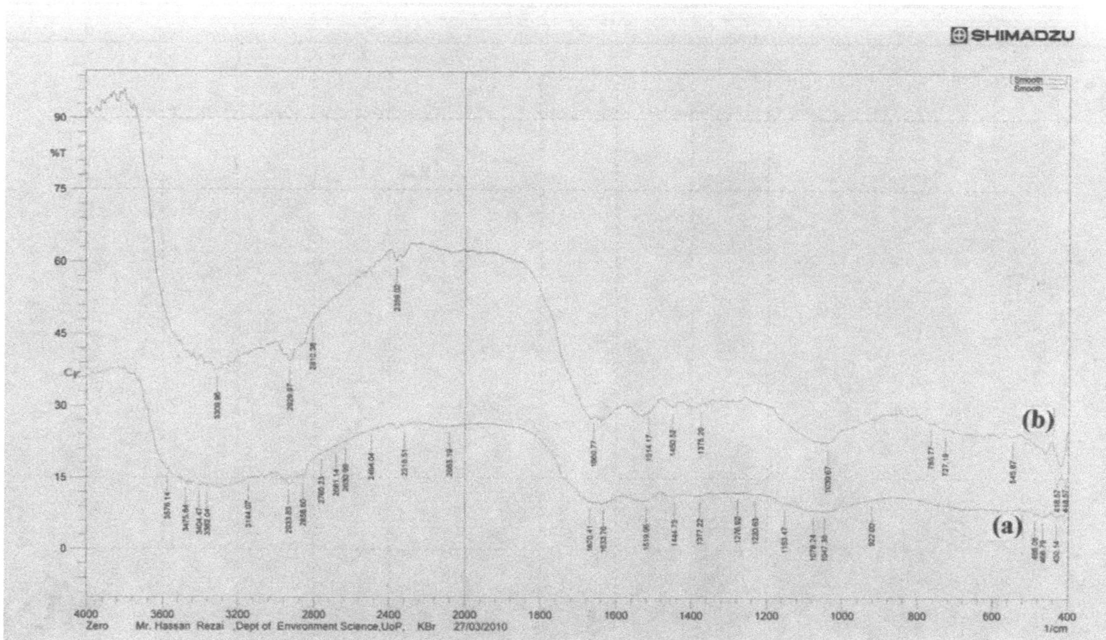

Fig.4.189 FTIR Spectrum of the *Oscillatoria sp.* (before **(b)** and after **(a)** biosorption of Cr).

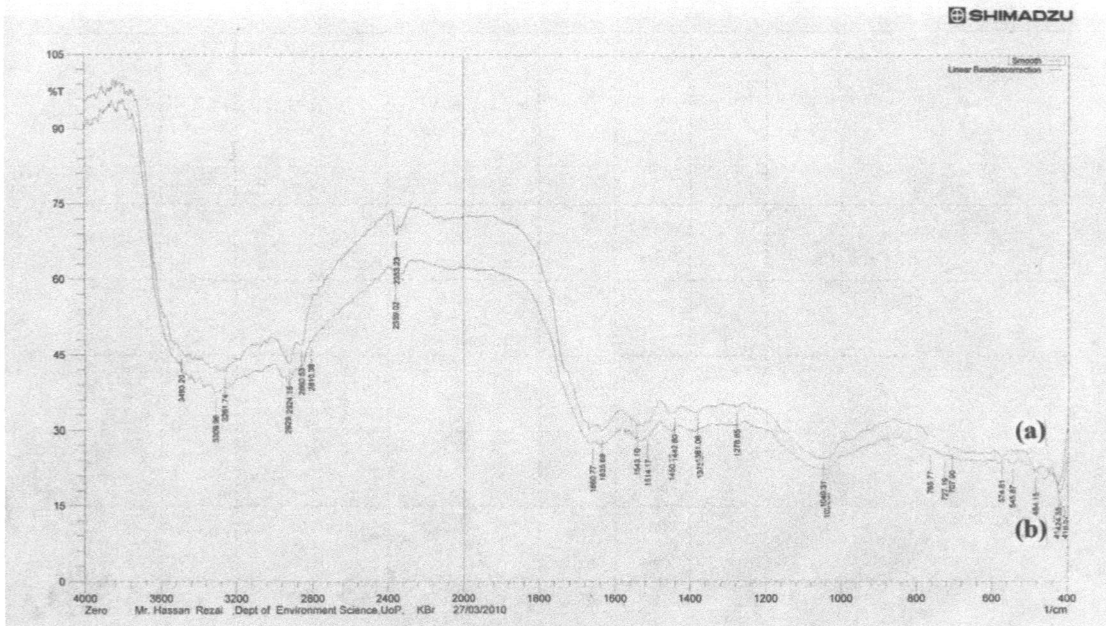

Fig.4.190 FTIR Spectrum of the *Oscillatoria sp.* (before **(b)** and after **(a)** biosorption of Cu).

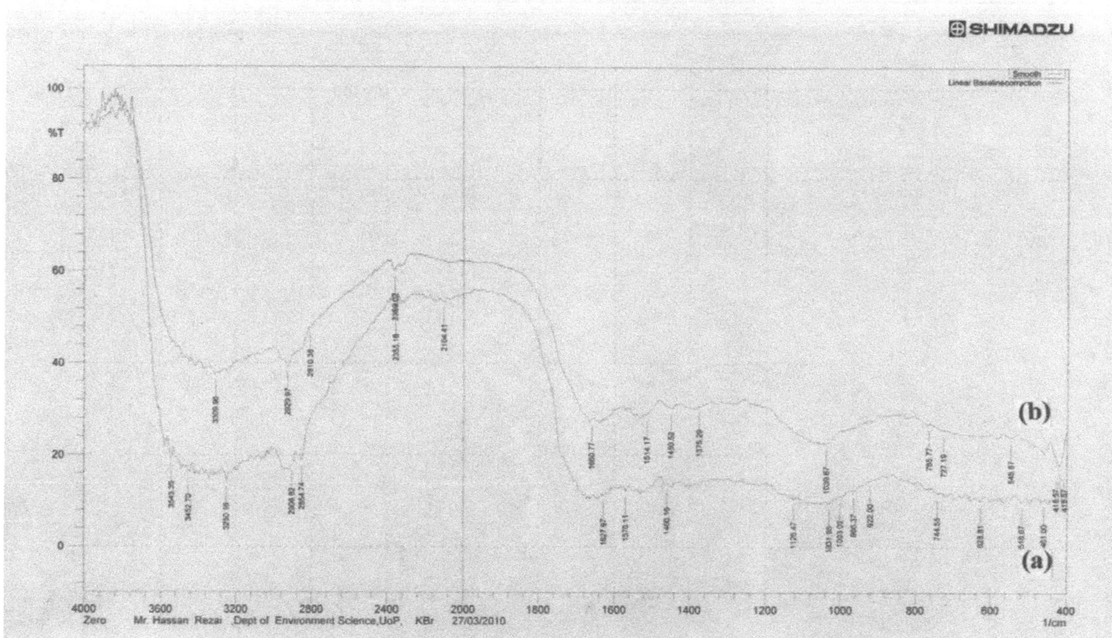

Fig.4.191 FTIR Spectrum of the *Oscillatoria sp.* (before **(b)** and after **(a)** biosorption of Ni).

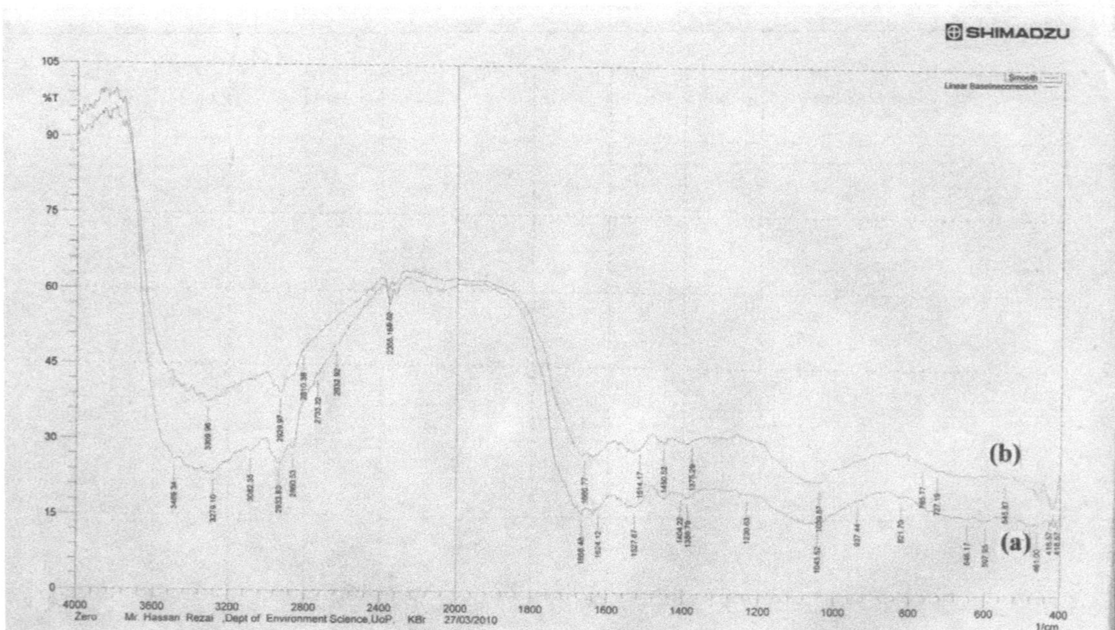

Fig.4.192 FTIR Spectrum of the *Oscillatoria sp.* (before **(b)** and after **(a)** biosorption of Zn).

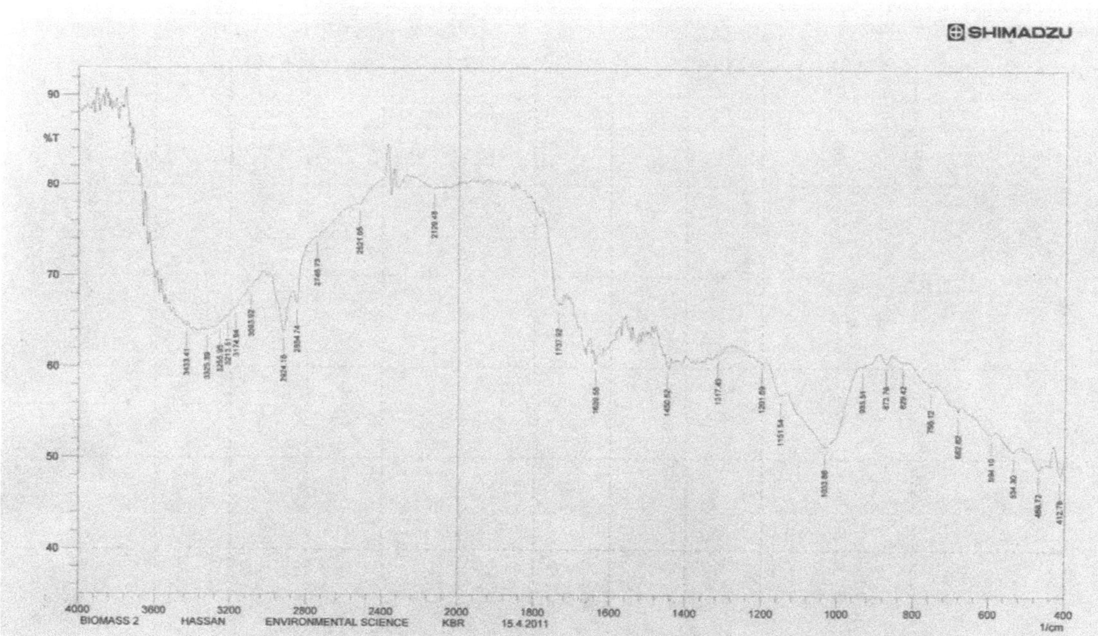

Fig.4.193 FTIR Spectrum of the *Spirulina sp.* (before biosorption).

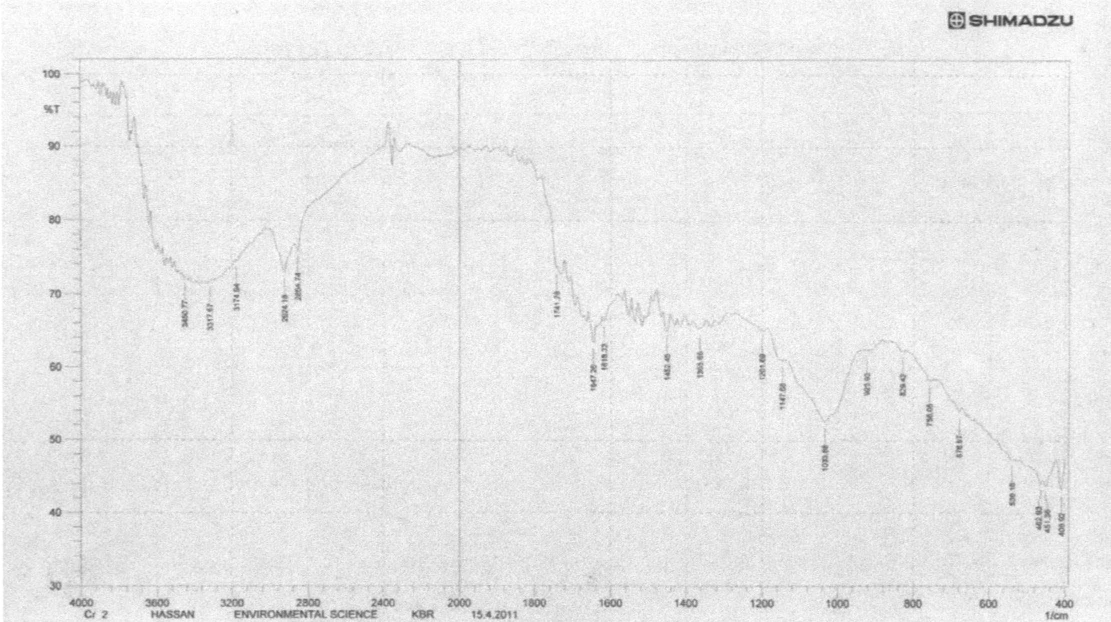

Fig.4.194 FTIR Spectrum of the *Spirulina sp.* (after biosorption of Cr).

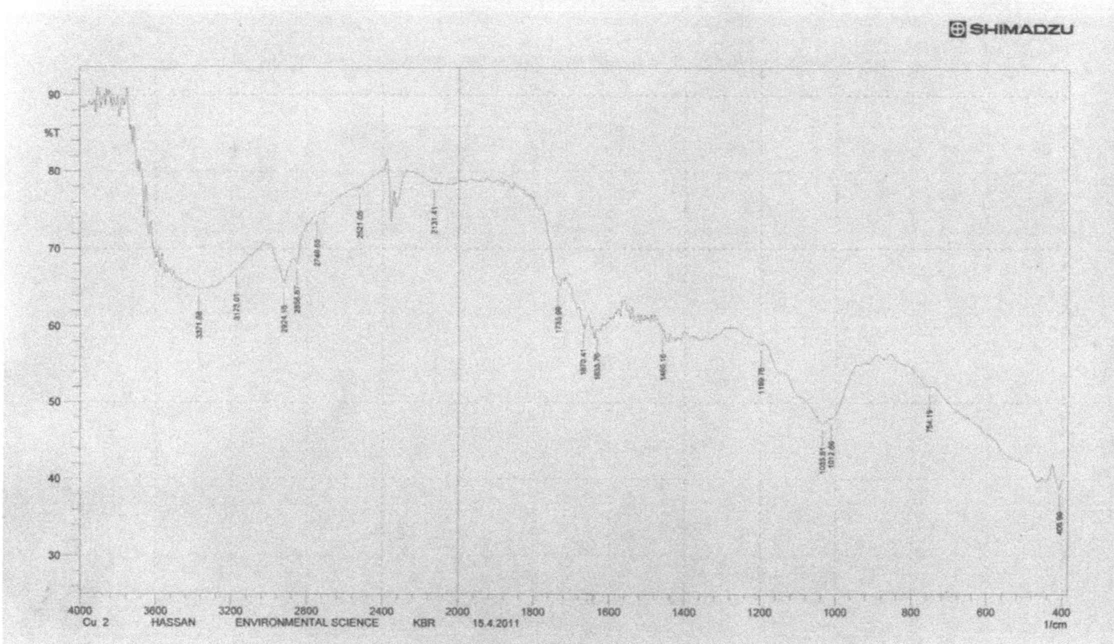

Fig.4.195 FTIR Spectrum of the *Spirulina sp.* (after biosorption of Cu).

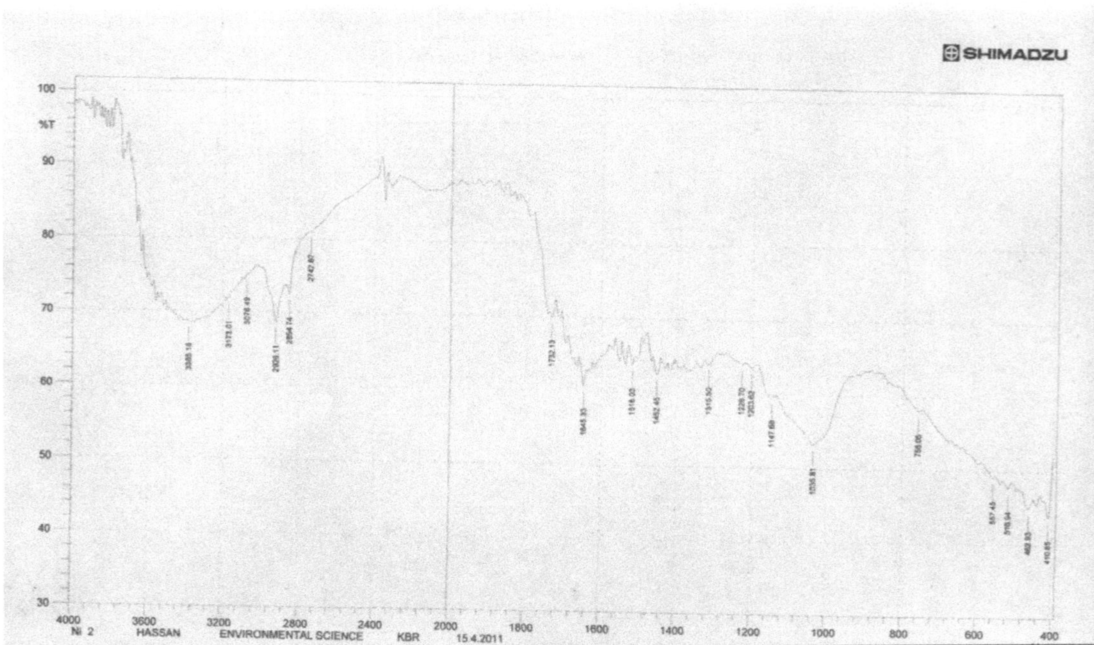

Fig.4.196 FTIR Spectrum of the *Spirulina sp.* (after biosorption of Ni).

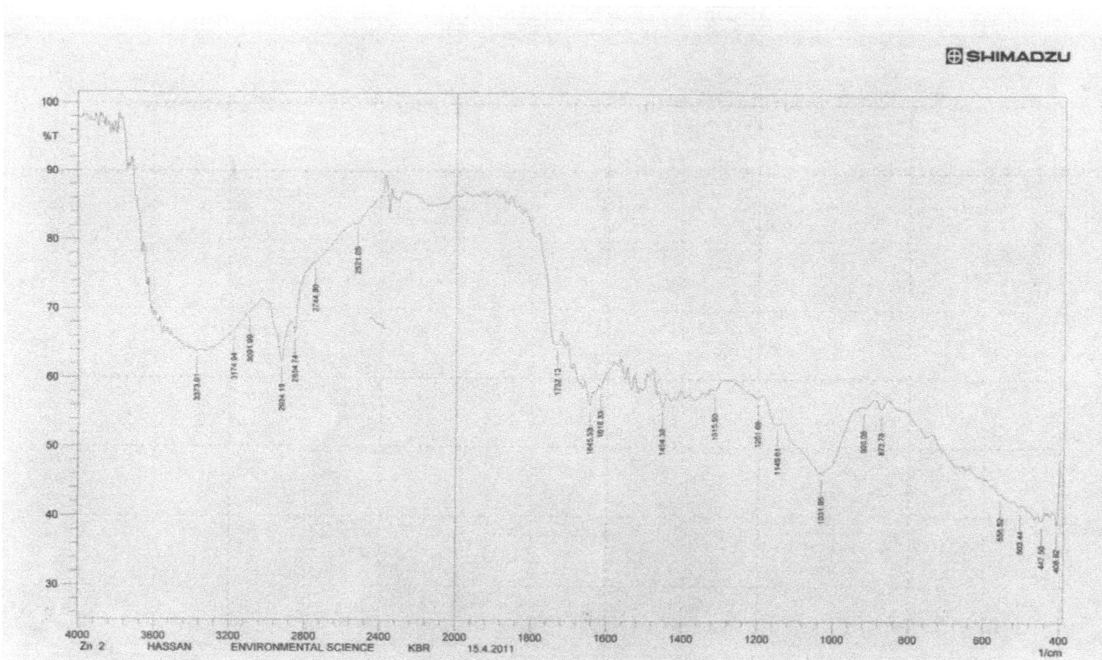

Fig.4.197 FTIR Spectrum of the *Spirulina sp.* (after biosorption of Zn).

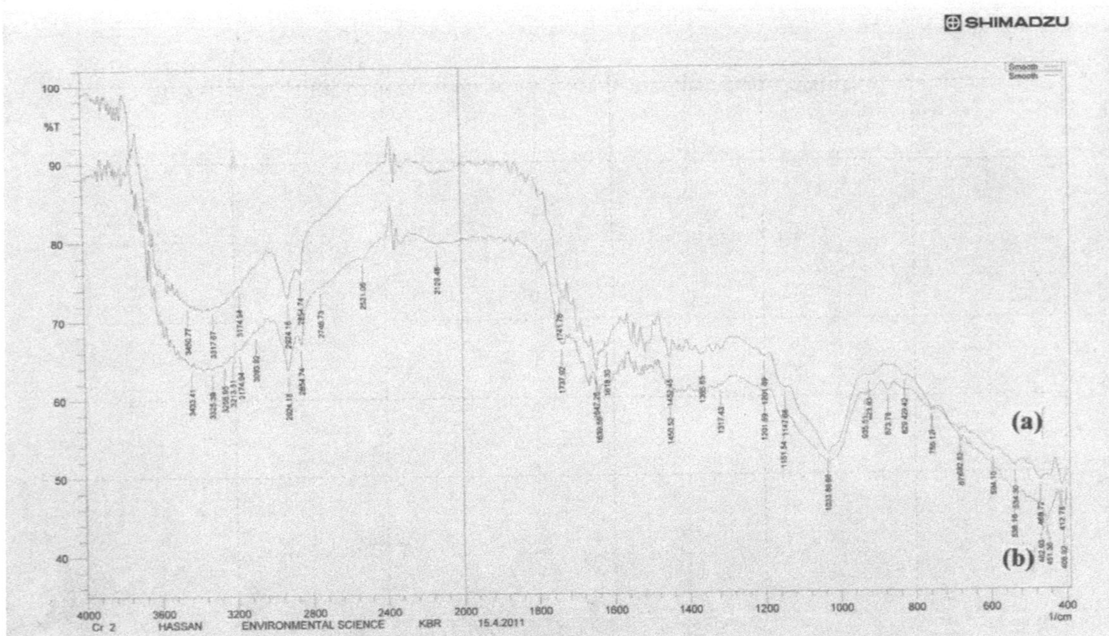

Fig.4.198 FTIR Spectrum of the *Spirulina sp.* (before **(b)** and after **(a)** biosorption of Cr).

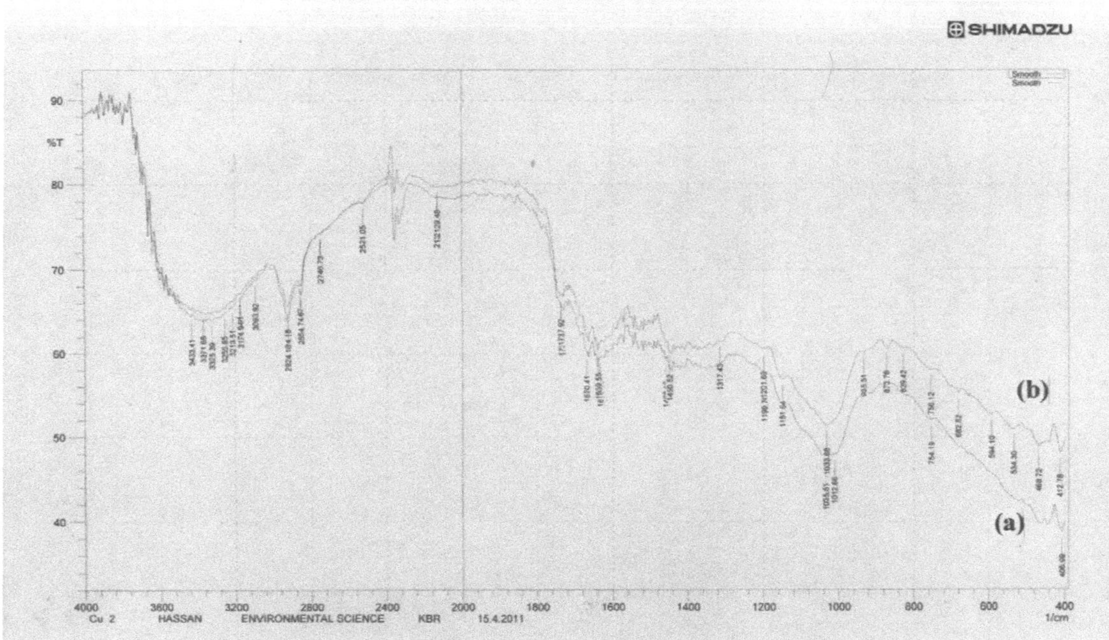

Fig.4.199 FTIR Spectrum of the *Spirulina sp.* (before **(b)** and after **(a)** biosorption of Cu).

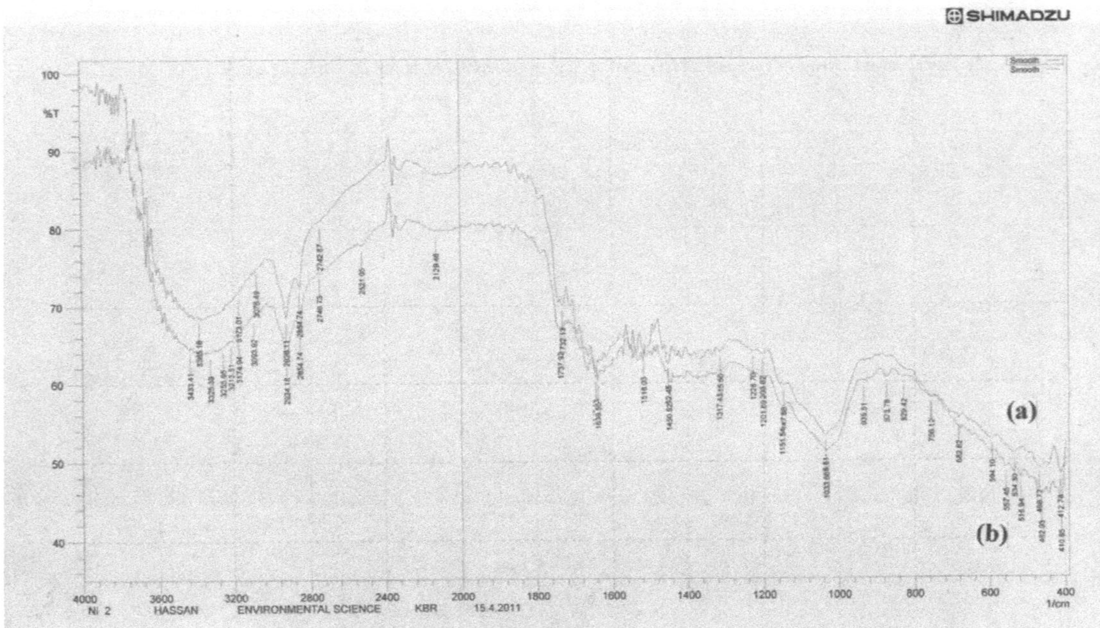

Fig.4.200 FTIR Spectrum of the *Spirulina sp.* (before **(b)** and after **(a)** biosorption of Ni).

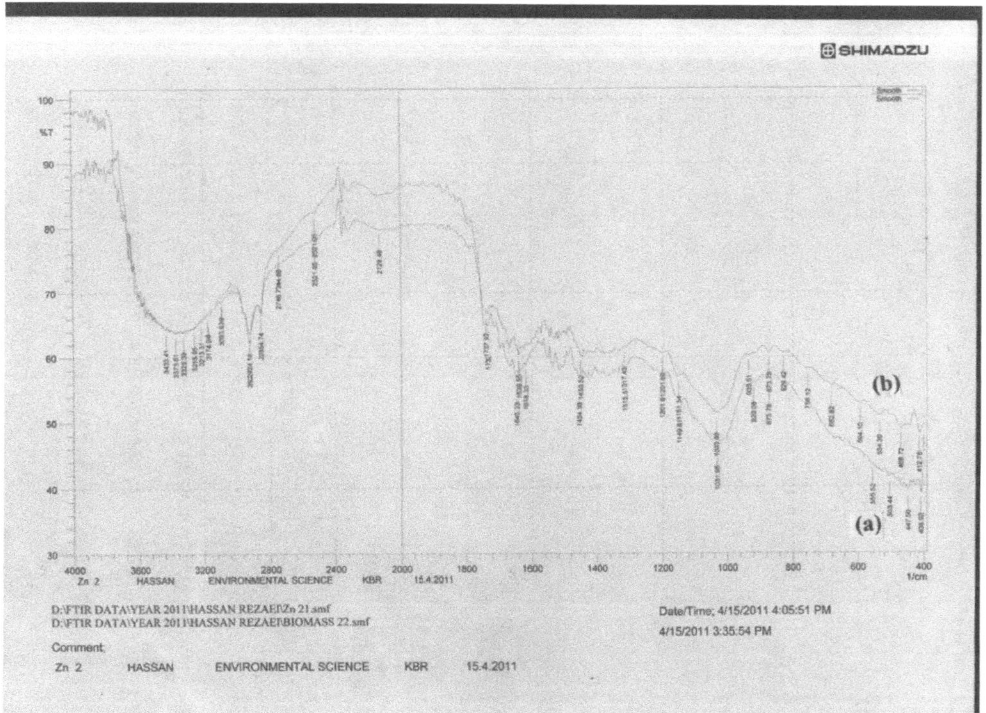

Fig.4.201 FTIR Spectrum of the *Spirulina sp.* (before **(b)** and after **(a)** biosorption of Zn).

4.34 Scanning electron microscopy (SEM)

Scanning electron microscopy (SEM) of *Chlorella pyrenoidosa, Oscillatoria sp.* and *Spirulina sp.* at before and after biosorption of chromium, copper, nickel and zinc present in the Figures 4.202 to 4.216.

The scanning electron micrograph clearly revealed the surface texture and morphology of the biosorbent at different magnifications. The SEM analysis revealed important information on surface morphology. In these micrographs structures with large surface area were evident. The results obviously show the difference between before and after loading of ions on the biomass surface.

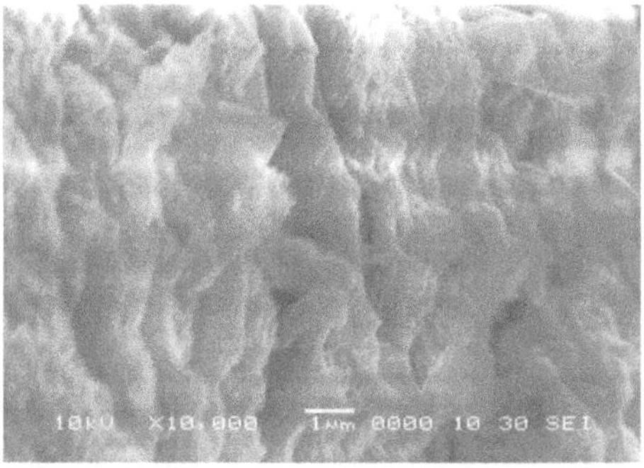

Fig.4.202 Scanning electron microscopy (SEM) micrographs of the *Chlorella pyrenoidosa* before biosorption.

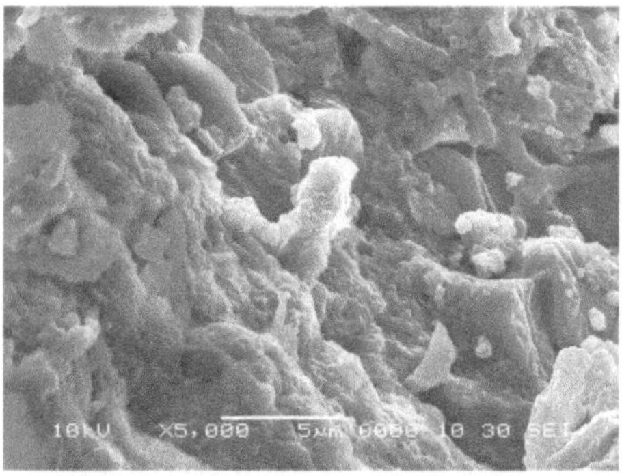

Fig.4.203 Scanning electron microscopy (SEM) micrographs of the *Chlorella pyrenoidosa* after biosorption of Cr.

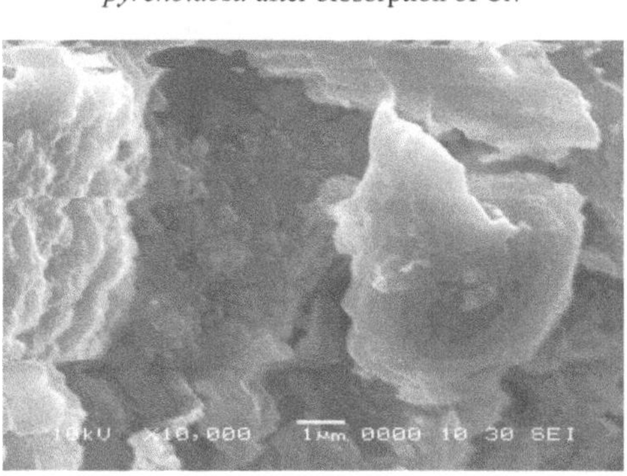

Fig.4.204 Scanning electron microscopy (SEM) micrographs of the *Chlorella pyrenoidosa* after biosorption of Cu.

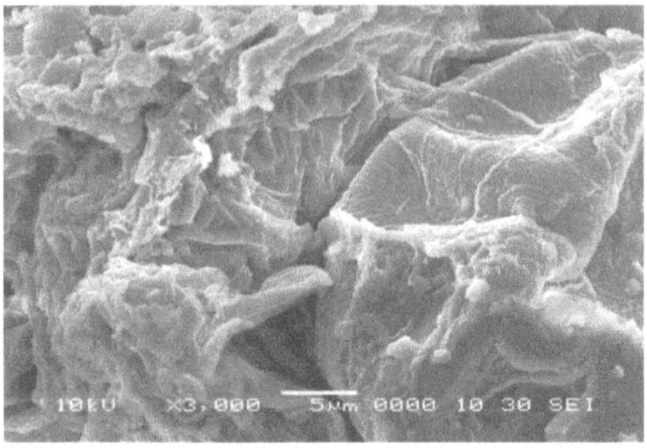

Fig.4.205 Scanning electron microscopy (SEM) micrographs of the *Chlorella pyrenoidosa* after biosorption of Ni.

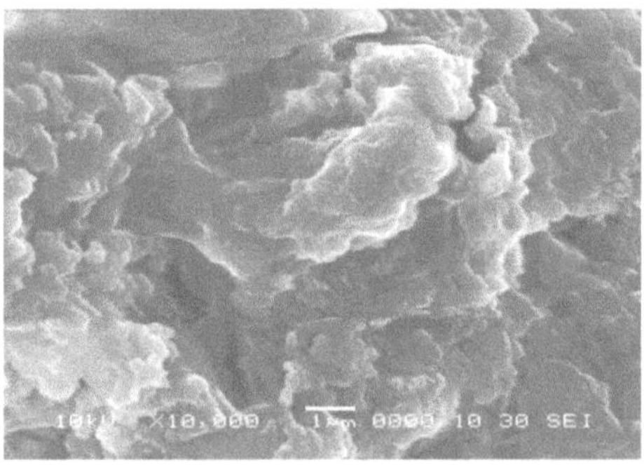

Fig.4.206 Scanning electron microscopy (SEM) micrographs of the *Chlorella pyrenoidosa* after biosorption of Zn.

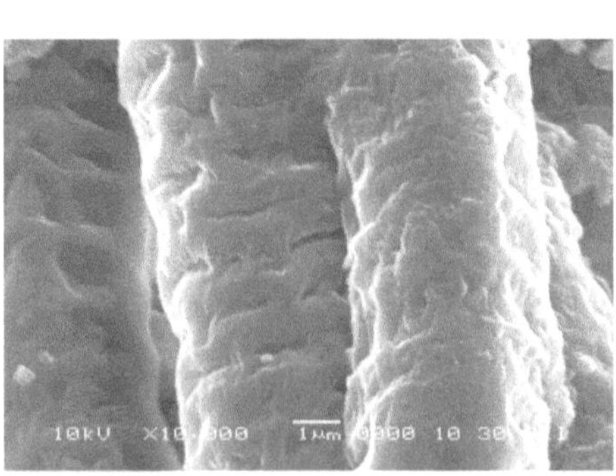

Fig.4.207 Scanning electron microscopy (SEM) micrographs of the *Oscillatoria sp.* before biosorption.

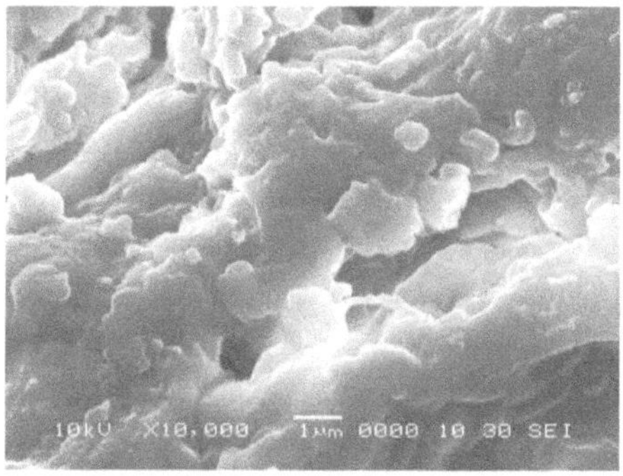

Fig.4.208 Scanning electron microscopy (SEM) micrographs of the *Oscillatoria sp.* after biosorption of Cr.

Fig.4.209 Scanning electron microscopy (SEM) micrographs of the *Oscillatoria sp.* after biosorption of Cu.

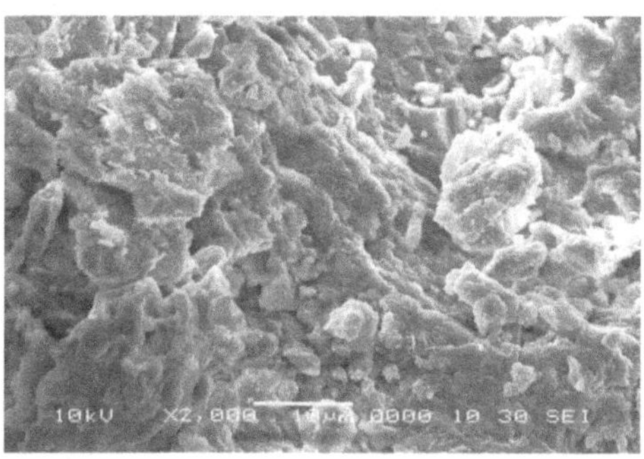

Fig.4.210 Scanning electron microscopy (SEM) micrographs of the *Oscillatoria sp.* after biosorption of Ni.

Fig.4.211 Scanning electron microscopy (SEM) micrographs of the *Oscillatoria sp.* after biosorption of Zn.

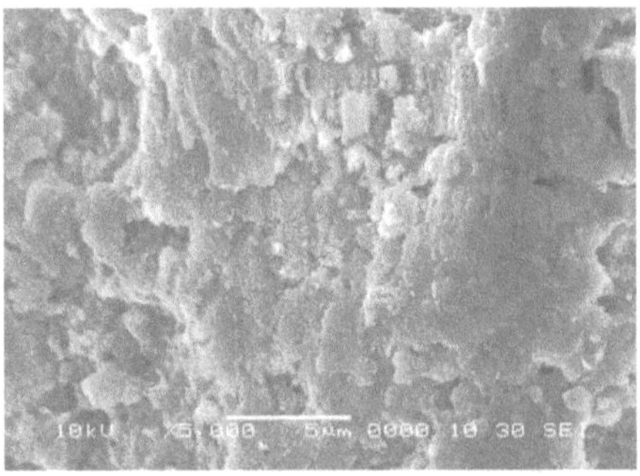

Fig.4.212 Scanning electron microscopy (SEM) micrographs of the *Spirulina sp.* before biosorption.

Fig.4.213 Scanning electron microscopy (SEM) micrographs of the *Spirulina sp.* after biosorption of Cr.

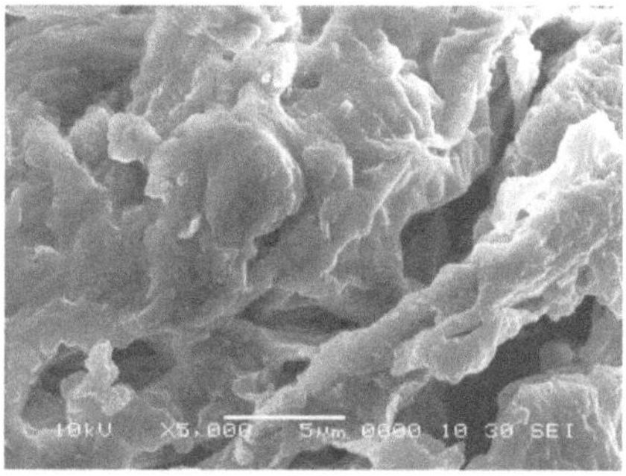

Fig.4.214 Scanning electron microscopy (SEM) micrographs of the *Spirulina sp.* after biosorption of Cu.

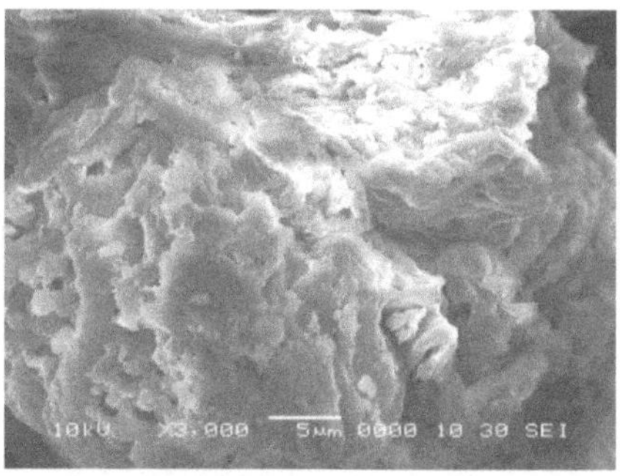

Fig.4.215 Scanning electron microscopy (SEM) micrographs of the *Spirulina sp.* after biosorption of Ni.

Fig.4.216 Scanning electron microscopy (SEM) micrographs of the *Spirulina sp.* after biosorption of Zn.

Chapter V: Summary and Conclusion

5.1 Summary

The potential of algaes (*Oscillatoria sp., Chlorella pyrenoidosa & Spirulina sp.*) for removing of heavy metals (Cr, Cu, Ni and Zn ions) was demonstrated through a series of batch and continuous column studies. Physical chemistry parameters such as concentration, pH, amount of biomass, temperature, adsorption isotherm, kinetic modeling, thermodynamic parameters and changes of functional groups and surface of biosorbent in the biosorption process by using of FTIR and SEM were studied.

In effect of time and initial concentration, percentage of remove of heavy metals and metal ion uptake capacity *for the three biomasses under study has been established as:*

Cr *Oscillatoria sp. > Chlorella pyrenoidosa > Spirulina sp.*
Cu *Oscillatoria sp. > Chlorella pyrenoidosa > Spirulina sp.*
Ni *Chlorella pyrenoidosa > Oscillatoria sp. > Spirulina sp.*
Zn *Oscillatoria sp. > Spirulina sp. > Chlorella pyrenoidosa*

When initial concentration increased to 100 mg/L:

Cr *Oscillatoria sp.* > *Chlorella pyrenoidosa* > *Spirulina sp.*

Cu *Oscillatoria sp.* > *Spirulina sp.* > *Chlorella pyrenoidosa*

Ni *Oscillatoria sp.* > *Spirulina sp.* > *Chlorella pyrenoidosa*

Zn *Oscillatoria sp.* > *Spirulina sp.* > *Chlorella pyrenoidosa*

In effect of pH and initial concentration of 10 mg/L, percentage of remove of heavy metals and metal ion uptake capacity *for the three biomasses under study has been established as:*

Cr *Oscillatoria sp.* > *Chlorella pyrenoidosa* > *Spirulina sp.*

Cu *Oscillatoria sp.* > *Spirulina sp.* > *Chlorella pyrenoidosa*

Ni *Chlorella pyrenoidosa* > *Spirulina sp.* > *Oscillatoria sp.*

Zn *Spirulina sp.* > *Chlorella pyrenoidosa* > *Oscillatoria sp.*

When initial concentration increased to 50 mg/L:

Cr *Chlorella pyrenoidosa* > *Oscillatoria sp.* > *Spirulina sp.*

Cu *Oscillatoria sp.* > *Spirulina sp.* > *Chlorella pyrenoidosa*

Ni *Chlorella pyrenoidosa* > *Spirulina sp.* > *Oscillatoria sp.*

Zn *Spirulina sp.* > *Oscillatoria sp.* > *Chlorella pyrenoidosa*

In effect of biomass dosage and initial concentration of 10 mg/L, percentage of remove of heavy metals and metal ion uptake capacity *for the three biomasses under study has been established as:*

Cr *Chlorella pyrenoidosa* > *Oscillatoria sp.* > *Spirulina sp.*

Cu *Spirulina sp.* > *Oscillatoria sp.* > *Chlorella pyrenoidosa*

Ni *Oscillatoria sp.* > *Chlorella pyrenoidosa* > *Spirulina sp.*

Zn *Spirulina sp.* > *Chlorella pyrenoidosa* > *Oscillatoria sp.*

When initial concentration increased to 50 mg/L,

Cr *Oscillatoria sp.* > *Chlorella pyrenoidosa* > *Spirulina sp.*

Cu *Spirulina sp.* > *Oscillatoria sp.* > *Chlorella pyrenoidosa*

Ni *Chlorella pyrenoidosa* > *Oscillatoria sp.* > *Spirulina sp.*

Zn *Oscillatoria sp.* > *Spirulina sp.* > *Chlorella pyrenoidosa*

In effect of temperature and initial concentration of 50 mg/L, percentage of remove of heavy metals and metal ion uptake capacity *for the three biomasses under study has been established as:*

Cr *Oscillatoria sp.* > *Chlorella pyrenoidosa* > *Spirulina sp.*

Cu *Spirulina sp.* > *Oscillatoria sp.* > *Chlorella pyrenoidosa*

Ni *Chlorella pyrenoidosa* > *Spirulina sp.* > *Oscillatoria sp.*

Zn *Oscillatoria sp.* > *Spirulina sp.* > *Chlorella pyrenoidosa*

The maximum ions uptake capacity (q_{max}) in Langmuir model for *the three biomasses under study for has been established as:*

Cr *Chlorella pyrenoidosa* > *Spirulina sp.* > *Oscillatoria sp.*

Cu *Oscillatoria sp.* > *Chlorella pyrenoidosa* > *Spirulina sp.*

Ni *Chlorella pyrenoidosa* > *Oscillatoria sp.* > Spirulina sp.

Zn *Oscillatoria sp.* > *Spirulina sp.* > *Chlorella pyrenoidosa*

The equilibrium experimental results of heavy metals have been fitted in the Langmuir and Freundlich models but coefficient of determination (R^2) Freundlich is bigger than of Langmuir and that means, Freundlich model is better model for our study. The value of 'R' is significant at 0.05 level according to the table given by Jog & Saptarshi (1980).

The R_L values at different initial adsorbate concentrations indicate favorable adsorption for all the adsorbents and adsorbates studied.

Parameter of kinetic model for heavy metals of biomasses showed that, *the* coefficient of determination (R^2) and K_2 HO'S Model for *the different metal ion concentration under study for has been established as:*

10 ppm > 20 ppm > 50 ppm > 100 ppm

With increase the initial concentration coefficient of determination (R^2) and K_2 decreased. The value of 'R' is significant at 0.05 level according to the table given by Jog & Saptarshi (1980).

A positive value of $\Delta H°$ indicated the endothermic nature of the process. A negative value of the free energy (ΔG°) indicated the spontaneous nature of the adsorption process. It was also noted that the change in free energy, increases with increase in temperature, which exhibits an increase in adsorption with rise in temperature. This could be possibly because of activation of more sites on the surface of biomasses with increase in temperature. A positive value of $\Delta S°$ showed increased randomness at solid–liquid interface during the adsorption of heavy metals, it also suggests some structural changes in the adsorbate and the adsorbent. Coefficients of determination (R^2) of *Chlorella pyrenoidosa* and *Spirulina sp.* indicate that the temperature has significant effect on the biosorption of heavy metals. The value of 'R' for *Chlorella pyrenoidosa* and *Spirulina sp.* is significant at 0.05 level according to the table given by Jog & Saptarshi (1980).The coefficients of determination (R^2) of *Oscillatoria sp.* indicate that the temperature has no significant effect on the biosorption of heavy metals. The value of 'R' is not significant at 0.05 level according to the table given by Jog & Saptarshi (1980).

Percentage of remove of 10 mg / L for Cr, Cu, Ni and Zn *for the three biomasses at breakthrough under study has been established as:*

Cr *Chlorella pyrenoidosa > Oscillatoria sp. > Spirulina sp.*

Cu *Spirulina sp. > Oscillatoria sp. > Chlorella pyrenoidosa*

Ni *Chlorella pyrenoidosa > Spirulina sp. > Oscillatoria sp.*

Zn *Spirulina sp. > Oscillatoria sp. >Chlorella pyrenoidosa*

In effect of time and initial concentration of 10 mg / L for Cr, Cu, Ni and Zn, percentage of remove of heavy metals and bed volume *for the biomasses under study has been established as:*

Chlorella pyrenoidosa	Ni > Cr > Cu > Zn
Oscillatoria sp.	Cr > Cu > Ni > Zn
Spirulina sp.	Zn > Cu > Ni > Cr

All of the result showed that the nitric acid was the best chemical solution for desorption process.HNO_3 > HCL > EDTA

FTIR Spectrums of *Chlorella pyrenoidosa, Oscillatoria sp.* and *Spirulina sp.* revealed the presence of hydroxyl, amino, carboxylic and carbonyl groups. The presence of OH group along with carbonyl group confirmed the presence of carboxylic acid groups in the biosorbent. The presence of NH group and OH group along with carbonyl group might be attributed the presence of amino acid groups in the biosorbent. The stretching vibration of OH, NH and C=O groups shifted for a certain extent in addition, which indicated that these three groups possibly involved in the biosorption. Assignment of bands to functional group on the surface of *Chlorella pyrenoidosa, Oscillatoria sp.* and *Spirulina sp.*.

Scanning electron microscopy (SEM) of *Chlorella pyrenoidosa, Oscillatoria sp.* and *Spirulina sp.* at before and after biosorption of chromium, copper, nickel and zinc clearly revealed the surface texture and morphology of the biosorbent at different magnifications.

5.2 Conclusions

1. All of three biomasses (*Chlorella pyrenoidosa, Oscillatoria sp.* and *Spirulina sp.*) has good potential for remove a heavy metals from waste water and they can be used in industrial wastewater treatment for remove the heavy metals.

2. With increase the initial concentration percentage of biosorption decreased and metal ion uptake capacity was increased.

3. With increase the pH up to 5 percentage of biosorption increased and metal ion uptake capacity was decreased and after pH=5 percentage of biosorption decreased and metal ion uptake capacity was increased

4. With increased the amount of biomass observed that percentage of biosorption increased and metal ion uptake capacity was decreased.

5. Optimum temperature and pH was 25 °C and 5 respectively.

6. The equilibrium experimental results of heavy metals have been fitted in the Langmuir and Freundlich models but coefficient of determination (R^2) Freundlich is bigger than of Langmuir and that means, Freundlich model is better model for our study.

7. Equilibrium time was about 60 minute.

8. The R_L values at different initial adsorbate concentrations indicate favourable adsorption for all the adsorbents and adsorbates studied.

9. Parameter of kinetic model for heavy metals of biomasses showed that; with increase the initial concentration coefficient of determination (R^2) and K_2 decreased.

10. A positive value of $\Delta H°$ indicated the endothermic nature of the process.

11. A negative value of the free energy ΔG° indicated the spontaneous nature of the adsorption process. It was also noted that the change in free energy, increases with increase in temperature, which exhibits an increase in adsorption with rise in temperature. This could be possibly because of activation of more sites on the surface of biomasses with increase in temperature.

12. A positive value of $\Delta S°$ showed increased randomness at solid–liquid interface during the adsorption of heavy metals, it also suggests some structural changes in the adsorbate and the adsorbent.

13. Spectrums of *Chlorella pyrenoidosa, Oscillatoria sp.* And *Spirulina* showed that revealed the presence of hydroxyl, amino, carboxylic and carbonyl groups. The presence of OH group along with carbonyl group confirmed the presence of carboxylic acid groups in the biosorbent. The presence of NH group and OH group along with carbonyl group might be attributed the presence of amino acid groups in the biosorbent.

14. Scanning electron microscopy (SEM) of *Chlorella pyrenoidosa, Oscillatoria sp.* and *Spirulina sp.* at before and after biosorption of chromium, copper, nickel and zinc clearly showed biosorption of heavy metals by algae biomass.

15. All of the result showed that the nitric acid was the best chemical solution for desorption process.

5.3 Recommendations for future work

- *Investigate the stability of the biomass as a function of the number of adsorption-desorption cycles.*
- *Evaluate the biosorption process for different types of wastewater of varying metal content.*
- *Investigate the potential of using the biosorption process as a remediation process for contaminated underground water.*
- *Establish a mathematical model that would link between the Langmuir model for batch tests and breakthrough curves in the fixed bed column.*

- Investigate the removal of heavy metals from industrial wastewaters using immobilized biomass beads.

- Evaluate the effectiveness of biomasses in the removal of organics compounds.

- *Evaluate the biosorption process for different types of wastewater of varying metal content by using fungi.*

REFERENCES

Abdel-Jawad, Al-Shammari and Al-Sulaimi, (2002): Non-conventional Treatment of treated municipal wastewater for reverse osmosis, *Desalination*, 142(1) 11-18.

Abdelkader Iddou, Mohamed Hadj Youcef, Abdallah Aziz,Mohand Said Ouali, (2011): Biosorptive removal of lead (II) ions from aqueous solutions using Cystoseira stricta biomass: Study of the surface modification effect, *Journal of Saudi Chemical Society* ,15, 83–88.

Abu Al-Rub F., Kandah M and Aldabaibeh N., (2002): Nickel removal from aqueous solutions using sheep manure wastes, *Eng. Life Sci.*2, 111–116.

Ahalya N, Kanamadi RD and Ramachandra TV, (2006): Biosorption of Iron (III) using the husk of Cicer arientinum. *Indian Journal of Chemical Technology*, 13, 122-127.

Ahalya N. and Ramachandra T.V. ,(2002): Restoration of wetlands - Feasibility Aspects of Biological Restoration presented at the National Conference on Aquatic Restoration and Biodiversity – Feb 15-16 in Kongunadu Arts and Science College, India.

Ahalya N, Ramachandra T.V., Kanamadi R.D, (2003): Biosorption of heavy metals, *Journal of Chemistry and Environment*, 7(4): 71-79.

Aksu, Z. and Kutsal, T., (1990): A comparative study for biosorptin characteristics of heavymetals ions with C. vzrlgaris. *Environmental Technology*, 21,979-987.

Al-Asheh S. and Duvnjak Z., (1997): Sorption of cadmium and other heavy metals by pine bark, Adv. *Environ. Res.* 1, 194.

Aldor, I., Fourest, E., and Volesky, B., *(1995): Desorption of cadmium from algal biosorbent. Can. J. Chem. Eng., 73, 516 - 522.*

Allen, S. J., Mckay b, G., & Porter, J. F., (2004): Adsorption isotherm models for basic dye adsorption by peat in single and binary component systems. *Journal of Colloid and Interface Science*, 280, 322–333.

Alpana Singh, Dhananjay Kumar, J.P. Gaur, (2008): Removal of Cu (II) and Pb (II) by Pithophora oedogonia: Sorption,desorption and repeated use of the biomass, *Journal of Hazardous Materials* ,152 , 1011–1019.

Anon, (1978): Heavy Metal Removal? Try starch xanthate. *Prod. Finishing*, 31, 72-74.

Anonymous, *(1989): Amberlite IRC-50. Philadelphia, PA, Rohm and Hass Co.,Inc.; Report number IE-232a.*

Aravindhan Rathinam *et al*., (2010): Biosorption of cadmium metal ion from simulated wastewaters using Hypnea valentiae biomass: A kinetic and thermodynamic study, *Bioresource Technology*, 101, 1466–1470.

Arun Dheetcha , Sandhya Mishra, (2008): Biosequestering Potential of Spirulina platensis for Uranium, *Curr Microbiol* ,57:508–514.

Atkinson, B. W., Bux, F. and Kasan, H. C., (1998): Considerations for application of biosorption technology to remedicrte metal-contaminated industriai effluents. *Water SA*, 24, 129-135.

Atkinson, B. W., Bux, F. and Kasan, H. C., (1996): Bioremediation of metal contaminateci industrial effluents using waste sludges. *Water Science and Technology*, 34, 9-15.

Aysegul Seker *et al*., (2008): Equilibrium, thermodynamic and kinetic studies for the biosorption of aqueous lead (II), cadmium (II) and nickel (II) ions on Spirulina platensis, Journal of Hazardous Materials 154, 973–980.

Bansal O.P., M. Prasad, S.N. Srivastava, (1982): Adsorption of oxamylon montmoril-lonites: a thermodynamic approach to the adsorption mechanism, *Agric. Biol. Chem*. 46 (1982) 323–332.

Beliles R.P., (1979): The lesser metals. In: F.W. Oehme, Editor, Toxicity of Heavy Metals in the Environment, Part 2, Marcel Dekker, New York, 383.

Beolchini *et al*., (2001): Modeling of Copper Biosorption by Arthrobacter sp. in a UF/MFMembrane Reactor, *Environ. Sci. Technol*, 35, 3048-3054.

Beveridge, T. J. and Koval, S. F., (1981): Binding of metals to cell envelops of Escherichia coli K- 12. *Applied Environmental Microbiology*, 42,325-335.

Brady, D., Stoll, A. and Duncan, J. R., (1994): Biosorption of heavy metals cations by nonviable yeast biomass. *Environmental Technology*, 15, 42948.

Brady, J. M. and Tobin, J. M., (1994): Adsorption of metal ions by Mitopus arrhims biomass: characterization studies. *Enzyme and Microbial Technology*, 6,671-675.

Brooks, C.S., *(1991): Metal recovery from industrial waste. Chelsea, MI, Lewis Publishers.*

Bux F. and Kasan, H. C., (1994): Comparison of selected methods for relative assessment of surface charge on waste sludge biomass. *Water SA*, 20, 73-76.

Case, O.P., (1974): Metallic recovery from wastewaters utilizing cementation, EPA-270/2-74-008, U.S. Washington DC.

Cem Gok , Dogukan Alkim Turkozu ,Sule Aytas ,(2011) :Removal of Th(IV) ions from aqueous solution using bi-functionalized algae-yeast biosorbent, *J Radioanal Nucl Chem* ,287:533–541.

Chatterjee S.K. I. Bhattacharjee, G. Chandra, (2010): Biosorption of heavy metals from industrial waste water by Geobacillus Thermodenitrificans . *Journal of Hazardous Materials*, 175, 117–125.

Chen, Y. H., Li, F. A., (2010): Kinetic study on removal of copper (II) using goethite and hematite nanophotocatalysts. *Journal of Colloid and Interface Science*, 347, 277–281.

Chojnackaa K, Chojnackib A and Górecka H., (2005): Biosorption of Cr^{3+}, Cd^{2+} and

Cu2+ ions by blue–green algae Spirulina sp.: kinetics, equilibrium and the mechanism of the process. *Chemosphere*, 59(1), 75-84.

Cristina M. Monteiro Paula M. L. Castro F. Xavier Malcata, (2009): Use of the microalga Scenedesmus obliquus to remove cadmium cations from aqueous solutions, *World J Microbiol Biotechnol*, 25:1573–1578.

Cristina M. Monteiro , Paula M. L. Castro ,F. Xavier Malcata, (2010): Cadmium Removal by Two Strains of Desmodesmus pleiomorphus Cells , *Water Air Soil Pollut* , 208:17–27.

Darnall, D.W., Greene, B. and Gradea-Torresday, J., (1988) :Gold Binding to algae in
BioHydrometallurgy: Proceedings Internat. Symp., Norris, P.R. and Kelly D.P.,(eds), Sci Technol Letters, Kew, Surrey, U.K., 487-98.

Davis, T.A., Volesky, B. and Mucci, A., *(2003): A review of the biochemistry of heavy metal biosorption by brown algae. Water Research, 37, 4311 -4330.*

Dhananjay Kumar, J.P. Gaur, (2011): Metal biosorption by two cyanobacterial mats in relation to pH, biomass concentration, pretreatment and reuse, *Bioresource Technology* 102, 2529–2535.

Donghee Park, Yeoung-Sang Yun, Ji Hye Jo, and Jong Moon Park, (2006): Biosorption Process for Treatment of Electroplating Wastewater Containing Cr(VI): Laboratory-Scale Feasibility Test, *Ind. Eng. Chem. Res*. 45, 5059-5065.

Elena Lourie & Vishwanath Patil & Elin Gjengedal, (2010): Efficient Purification of Heavy-Metal-Contaminated Water by Microalgae-Activated Pine Bark, *Water Air Soil Pollut,* 210:493–500.

Elisabetta Finocchio, Alessandra Lodi, Carlo Solisio, Attilio Converti, (2010) :Chromium (VI) removal by methylated biomass of Spirulina platensis: The effect of methylation process, *Chemical Engineering Journal*, 156, 264–269.

EPA, (1980): Development document for effluent limitation guidelines and standards for the inorganic chemicals manufacturing point source category, USEPA, Effluent Guidelines Division, Office of Water and Waste Management, Washington, DC.

Esposito A., Pagnanelli F., Beolchini F., Dovi V. and Veglio F., *(2001): Cadmium and copper biosorption on Sphaerotilusnatans: influence of pH and biomass concentration on the biosorption modeling.* *Hydrometallurgy, 60, 129 -141.*

Figueira, MM, Yang, J, Volesky, B & Camargos, ERS., (1995): Interference of Fe in the Cd uptake by Sargassum biomass. Biohydrometallurgical Processing, Vol.2, Proc. Internat. Biohydromet. Symp., Jerez, CA, Vargas, T, Toledo, H & Wiertz,JV (eds.) Univ. Chile, Santiago, Chile, 187-194.

Figueira, Volesky, Mathieu, (1999): Instrumental Analysis Study of Iron Species Biosorption by Sargassum Biomass, *Environ. Sci. Technol.* 33, 1840-1846.

Fourest, E. and Roux, J.C., *(1994): Improvement of heavy metal biosorption by mycelial dead biomass* (Rhizopus arrhizus, Mucor miehei and Penicillium chrysogenum)*: pH control and cationic activation.* FEMS Microbial. *Rev. 14, 325 - 332.*

Freundlich H., (1906) :*Phys Chemie*, 57, 384.

Gadd, G. M., (1992): Microbial control of heavy metal pollution. in Microbioal Control of Pollution, eds. J. C. Fry, G. M. Gadd, R A. Herbert, C. W. Jones and 1. A. Watson-Craik, Cambridge University Press.

Gadd, G. M., (1988): Accumulation of metals by microorganisms and algae. In Biotechnology – A Comprehensive Treatise in 8 Volumes, eds. J. Rem and G. Reed, 68, VCH Publishers,Weinhheim, pp. 401-403.

Gadd, G.M. Fry, J.C, and Herbert, R. A., *(1992): Microbial control of pollution. In 48th Symposium of the Society for General Microbiology. University of Cardiff. Cambridge University Press, 59 - 89.*

Gadd GM. and White C., (1993): Microbial treatment of metal pollution - A working biotechnology. TIBTech.11, 353-359.

Gee, A.R. and Dudeney, A.W.L., (1988): Adsorption and crystallisation of gold atbiological surfaces, in BioHydrometallurgy: Proceedings Internat. Symp., Norris,P.R. and Kelly D.P., (eds), Sci Technol Letters, Kew, Surrey, U.K., 437-451.

Greene, B., McPherson, R. and Darnall, D., *(1987): Algal sorbents for selective metal ion recovery, in Metals Speciation, Separation and Recovery, Lewis, Chelsea, Ml, 315-338.*

Groffman, A., Peterson, S and Brookins, D., (1992): Removing lead from wastewater using zeolite. *Wat. Environ. Technol.*, 4, 54-59.

Guibal, E., Roulph, C. and Le Cloirec, P., *(1992): Uranium biosorption by a filamentous fungus* Mucor meihei: *pH effect on the mechanisms and performances of uptake.* *Water Res. 26, 1139 - 1149.*

Gupta, R., Ahuja, P., Khan, S., Saxena, R. K., and Mohapatra, H., *(2000): Microbial biosorbents: Meeting challenges of heavy metal pollution in aqueous solutions.* *Current Science, 78, 967 - 973.*

Gupta V. K., Shrivastava A. K. and Neeraj Jain., (2001): Biosorption of Chromium(VI) From Aqueous solutions by green algae spirogyra species. *Water Research.* 35 (17), 4079-408.

Gupta Vinod K. Arshi Rastogi, Arunima Nayak, (2010): Biosorption of nickel onto treated alga (Oedogonium hatei): Application of isotherm and kinetic models, *Journal of Colloid and Interface Science*, 342, 533–539.

Hammaini A., Ballester A., Gonzalez F., Blazquez M. L., and Munoz J. A., *(1999): Activated Sludge as Biosorbent of Heavy Metals. Biohydrometallurgy and the environment toward the mining of the 21st century. International biohydrometallurgy symposium IBS'99. 185- 192.*

Hancock, 1. C., (1986): Trace metal removal h m aqueous solution. in Trace Metal Removal from Aqueous Solution, eds. R. Thompson, Special Publication No. 61, The Royal Society of Chemistry, London, U.K., 25-43.

Harland, C.E., *(1994): Ion exchange: theory and practice. The Royal Society of Chemistry, Cambridge.*

Herrero Roberto *et al.,* (2005): Removal of inorganic mercury from aqueous solutionsby biomass of the marine macroalga *Cystoseira baccata .Water Res,* 39(14), 3199-3210.

Hikmet Katırcıoglu *et al.,* (2008): Removal of cadmium (II) ion from aqueous system by dry biomass, immobilized live and heat-inactivated Oscillatoria sp.H1 isolated from freshwater (Mogan Lake), *Bioresource Technology*, 99, 4185–4191.

Hiren Doshi, Arabinda Ray,.L.Kothari, Bharat Gami, (2006): Spectroscopic and Scanning Electron Microscopy Studies of Bioaccumulation of Pollutants by Algae, *Current Microbiology,* 53 ,148–157.

Hiren Doshi, Arabinda Ray, I.L. Kothari, (2007): Bioremediation Potential of Live and Dead Spirulina: Spectroscopic, Kineticsand SEM Studies. *Biotechnology and Bioengineering*, 96(2) 1051- 1063.

Ho Y.S., D.A.J. Wase, C.F. Forster, (1996): Kinetic studies of competitive heavy metal
adsorption by sphagnum moss peat, *Environ. Technol.* 17, 71–77.

Holan, Z.R. and Volesky, B, 1995: Accumulation of cadmium, lead and nickel by fungal and wood biosorbents. *Appl. Biochem. Biotechnol.* 53, 133-146.

Holan, Z.R. and Volesky, B., (1994): Biosorption of lead and nickel by biomass of marine algae. *Biotech Bioeng.* 43, 1001-9.

Holan, Z.R., Volesky, B., and Prasetyo, I., *(1993): Biosorption of cadmium by biomass of marine algae. Biotechnology and Bioengineering, 41, 819 - 825.*

Hughes, M. N. and Poole, R. K., (1989): Metals and Microorganisms, Chapman and Hall, London, 10.

Jog, Saptarshi, (1980): Statistical Methods.

Kadirvelu K., (1998): Preparation and characterization of activated carbon, from coir pith and its application to metal bearing wastewater, Ph.D. Thesis, Bharathiar University, Coimbatore, India.

Khoo K. M. & Ting Y. P., (2001): Biosorption of gold by immobilized fungal biomass, *Biochem Eng. J.*, 8, 51-59

Khovrychev, M. P., (1973): Adsorption of copper ions by cells of Cadida utilis, *Microbiology*, 42.755-749.

Kleinubing S.J., Silva E.A., Silva M.G.C., E. Guibal ,(2011): Equilibrium of Cu(II) and Ni(II) biosorption by marine alga Sargassum filipendula in a dynamic system: Competitiveness and selectivity, Bioresource Technology 102 4610–4617.

Klimmekadh *et al,* (2001): Comparative Analysis of the Biosorption of Cadmium, Lead,
Nickel, and Zinc by Algae, *Environ. Sci. Technol.* 35, 4283-4288.

Knorr D., (1991): Recovery and utilization of chitin and chitosan in food processing waste management, *Food Technol*, 45, 114-122.

Ku, Y.P., *(1987): Innovative uses for carbon adsorption of heavy metals from plating waste waters: 1. Activated carbon polishing treatment. Environmental Progress. 6, 119-124.*

Kuyicak N, Volesky B., (1990): Biosorption by fungal biomass. In: Volesky B, editor. Biosorption of heavy metals. Florida: CRC press; 173–98.

Kuyucak, N. and Volesky, B., *(1989): Accumulation of gold by algal biosorbent. Biorecovery, 1, 189 - 204.*

Kuyucak, N. and Volesky, B., (1988): New algal biosorbent for a gold recovery process in BioHydrometallurgy: Proceeding of the Internat. Symp.,Norris, P.R.and Kelly D.P., (eds), Sci Technol Letters, Kew, Surrey, U.K., 453-464.

Langmuir I., (1918): *J Amer Chem Soc*, 40, 1361.

Leila Chebil Ajjabi , Lassaad Chouba, (2009): Biosorption of Cu^{+2} and Zn^{+2} from aqueous solutions by dried marine green macroalga Chaetomorpha linum , *Journal of Environmental Management* 90 , 3485–3489.

Macaskie L.E., Empson, R.M., Cheetham, A.K., Grey, C.P. and Skarnaulis, A.J., (1992): Uranium bioaccumulation by a Citrobacter sp. as a result of enzymatically mediated growth of polycrystalline HUO_2PO_4. Science 257, 782-84.

Marcia Teresinha *et al.*,(2009): Biosorption of nickel (II) ions by using chemically pre-treated Sargassum filipendula biomass in a fixed bed column, *World J Microbiol Biotechnol*, 25:1849–1856.

Marta Izquierdo *et al.*, (2010): Modeling of copper fixed-bed biosorption from wastewater by Posidonia oceanic, *Bioresource Technology*, 101, 510–517.

Mattuschka, B., Junghaus, K. and Straube G., (1993): Biosorption of metals by waste
Biomass in Biohydrometallurgical Technologies. Proceedings of the internationalbiohydrometallurgical symposium, Torma, A.E., Apel, M.L. and Brierley, C.L.,(eds), The Minerals, Metals and Materials Society, Warrendale, PA.

McKay, G., Blair, H. S., & Gardner, J. R., (1982): Adsorption of dyes on chitin equilibrium studies. *Journal of Applied Polymer Science*, 27, 3043–3057.

Mofa A. S., (1955): Plants proving their worth in toxic metal cleanup, *Science*, 269,302-305.

Monica Patron-Prado *et al.*, (2010): Copper and Cadmium Biosorption by Dried Seaweed Sargassum sinicola in Saline Wastewater , *Water Air Soil Pollut* , 210:197–202.

Monica Patron-Prado *et al.*, (2011): Biosorption Capacity for Cadmium of Brown Seaweed Sargassum sinicola and Sargassum lapazeanum in the Gulf of California, *Water Air Soil Pollut.*

Montgomery J. M., *(1985): Water Treatment Principles and Design. John Wiley and Sons Inc. USA.*

Munleedharan, T. R. and Venkobachar, L., (1994): Further insight into the mechanism of biosorption of heavy metals by Ganoderma Lucidum. *Environmental Technology*, 15, 10 15-1027.

Mustafa Tuzen, Ahmet Sarı ,(2010): Biosorption of selenium from aqueous solution by green algae (Cladophora hutchinsiae) biomass: Equilibrium, thermodynamic and kinetic studies, *Chemical Engineering Journal* ,158 , 200–206.

Naimo Teresa J., (1995): A review of the effects of heavy metals on freshwater mussels, *Ecotoxicology*, Vol 4, No 6, pp 341 – 362.

Nirupama Mallick, (2003): Biotechnological potential of Chlorella vulgaris for accumulation of Cu and Ni from single and binary metal solutions, *World Journal of Microbiology & Biotechnology* 19: 695–701.

Pagnanelli *et al.,* (2000): Biosorption of Metal Ions on Arthrobacter sp.: Biomass Characterization and Biosorption Modeling, *Environ. Sci. Technol.*, 34, 2773-2778.

Pahlavanzadeh H. A.R. Keshtkarb, J. Safdarib, Z. Abadi, (2010): Biosorption of nickel (II) from aqueous solution by brown algae: Equilibrium, dynamic and thermodynamic studies, *Journal of Hazardous Materials,* 175 304–310.

Patterson, J.W., (1975): Wastewater Treatment Technology. Ann Arbor Science, Ann Arbor, Michigan.

Paul Chen and Lei Yang, (2005): Chemical Modification of Sargassum sp. for Prevention of Organic Leaching and Enhancement of Uptake during Metal Biosorption, *Ind. Eng. Chem. Res.* 44, 9931-9942.

Paul Chen, Lei Yang, (2006): Study of a Heavy Metal Biosorption onto Raw and Chemically Modified Sargassum sp. via Spectroscopic and Modeling Analysis, *Langmuir*, 22, 8906-8914.

Ping Xin Sheng, Yen-Peng Ting, and J. Paul Chen ,(2007): Biosorption of Heavy Metal Ions (Pb, Cu, and Cd) from Aqueous Solutions by the Marine Alga Sargassum sp. in Single- and Multiple-Metal Systems, *Ind. Eng. Chem. Res.* 46, 2438-2444.

Pradhan, A. A. and Levine, A. D., (1992): Role of extracellular components in microbial biosorption of copper and lead. *Water Science and Technology*, 26, 2153-2 156.

Prescott LM, Harley JP, Klein DA., (2005): Microbiology. McGraw-Hill Science/Engineering/Math. ed Fifth.

Rincon J, Gonzalez F, Ballester A, Blazquez ML, Munoz JA., (2005):Biosorption of heavy metals by chemically-activated alga Fucus vesiculosus. *J Chem Technol Biotechnol*; 80:1403–7.

Roberto Herrero *et al,*. (2011): Full description of copper uptake by algal biomass combining equilibrium NICA model with a kinetic intraparticle diffusion driving force approach, *Bioresource Technology* ,102 ,(2011) 2990–2997.

Robson C. Oliveira, Claire Jouannin, Eric Guibal, Oswaldo Garcia Jr., (2011): Samarium (III) and praseodymium (III) biosorption on Sargassum sp.: Batch study, Process Biochemistry 46 736–744.

Rodrigues Silva, A. C. Melo Ferreira, A. C. Costa, (2009): Uranium biosorption under dynamic conditions: Preliminary tests with Sargassum filipendula in real radioactive wastewater containing Ba, Cr, Fe, Mn, Pb, Ca and Mg, *Journal of Radioanalytical and Nuclear Chemistry*, 279(3), 909–914.

Romera E, Gonzalez F, Ballester A, Blazquez ML, Munoz JA., (2006): Biosorption with algae: a Statistical review. Crit Rev Biotechnol; 26:223–35.

Sayler G. S., Nelson J. D., Jr., and. Colwell R. R., (1975): Role of Bacteria in bioaccumulation of mercury in the Oyster Crassostrea virginica. *App. Microbial.*; 30(1): 91–96.

Shaik Basha and Bhavanath Jha, (2008): Estimation of Isotherm Parameters for Biosorption of Cd (II) and Pb(II) onto Brown Seaweed, Lobophora Wariegata, *J. Chem. Eng.* Data, 53, 449–455.

Sheng Ping Xin, Yen-Peng Ting, J. Paul Chen and Liang Hong, (2004): Sorption of lead, copper, cadmium, zinc, and nickel by marine algal biomass: characterization of biosorptive capacity and investigation of mechanisms. *Journal of Colloid and Interface Science*, 275 (1), 131-141.

Sigg, L., (1987): Surface chemical aspects of the distribution and fate of metal ions in lakes. In Aquatic Surfuce Chemistry, eds. Moo-Young Murray, Pergamon Press, New York, 235-247.

Siew-Ling Hii & Su-Yee Yong & Ching-Lee Wong, (2009): Removal of rhodamine B from aqueous solution by sorption on Turbinaria conoides (Phaeophyta), *J Appl Phycol*, 21:625–631.

Sitting M., (1976): Toxic Metals—Pollution Control and Worker Protection, Noyes Data Corporation, New Jersey.

Tobin, J. M., Cooper, D. G. and Neufeld, R. J., (1984): Uptake of metal ions by Rhizoprrs arrhizuz biomass. *Applied Environmental Microbiology*, 47, 821-824.

Treen-Sears, M.E., Volesky, B. and Neufeld, R.J., *(1984): Ion exchange/ complexation of the uranyl ion by* Rhizopus *biosorbent. Biotechnol. Bioeng. 26, 1323- 1329.*

Tuzun I, Bayramoglu G, Yalcin E, Basaran G, Celik G, Arica MY.,(2005): Equilibrium and kinetic studies on biosorption of Hg(II), Cd(II) and Pb(II) ions onto microalgae Chlamydomonas reinhardtii. *J Environ Manage*; 77:85–92.

U.S. EPA, (1979): Environmental Pollution: Control Alternatives: Economics of Wastewater Treatment Alternatives for the Electroplating Industry. EPA – 625/5- 79-016, Cincinnati, Ohio.

U.S.EPA., (1979a): Control Technology for the metal finishing industry – Evaporators, EPA-625/8-79-002, Cincinnati, Ohio.

Van der Heen, P., (1977): The removal of traces of heavy metals from drinking water and industrial effluent with ion exchanger. The Regional Chemical Society Meeting.

Veglio F, Beolchini F., (1997): Removal of metals by biosorption: *a review. Hydrometallurgy*, 44:301–16.

Veglio, Beolchini, Toro, (1998): Kinetic Modeling of Copper Biosorption by Immobilized Biomass, *Ind. Eng. Chem. Res*. 37, 1107-1111.

Vijayaraghavan K., Jegan J., Palanivelu K. and Velan M., (2005): Biosorption of copper, cobalt and nickel by marine green alga *Ulva reticulata* in a packed column. *Chemosphere* 60(3), 419-426.

Vijayaraghavan K, Yun YS., (2008): Bacterial Biosorbents and biosorption. *Biotechnol Adv*; 26:266–91.

Vijayaraghavan, Sathishkumar, Balasubramanian ,(2010): Biosorption of Lanthanum, Cerium, Europium, and Ytterbium by a Brown Marine Alga, Turbinaria Conoides, *Ind. Eng. Chem. Res*. 49, 4405–4411.

Volesky B., *(1990): Biosorption of Heavy Metals, CRC Press Inc, Boca Raton,Florida.*

Volesky B., (1990 a): Biosorption by fungal biomass. In: Volesky B, editor. Biosorption of heavy metals. Florida: CRC press; 139–71.

Volesky B., (1990 b): Introduction. In: Volesky B, editor. Biosorption of heavy metals. Boca Raton: CRC press; 3–5.

Volesky B., (1990 c): Removal and recovery of heavy metals by biosorption. In: Volesky B, editor. Biosorption of heavy metals. Florida: CRC press; 8-43.

Volesky B., *(1999): Biosorption for the Next Century. Biohydrometallurgy and the environment toward the mining of the 21st century. International biohydrometallurgy symposium IBS'99. 161 -170.*

Volesky, B., (1986): Biosorbent Materials, *Biotechnoi. Bioeng Symp.*, 16: 121-126.

Volesky B, Holan ZR. , (1995): Biosorption of heavy metals. *Biotechnol Prog*; 11:235–50.

Volesky, B., and Kuyucak, N., (1988): Biosorbent for gold, US Patent No. 4,769,223.

Volesky B., (2007): Biosorption and me. *Water Res*; 41:4017–29.

Wang JL, Chen C., (2006): Biosorption of heavy metals by Saccharomyces cerevisiae: *a review. Biotechnol Adv*; 24:427–51.

Watonabe, T and Ogawa, K., (1929): Activated carbon for purifying copper electrolytes. *Chem. Abs.* 24: 1037.

Yan Gao , Xue Yan,(2011): Response of Chara globularis and Hydrodictyon reticulatum to lead pollution: their survival, bioaccumulation and defense, *J Appl Phycol.*

Yi-Chao Lee, Shui-Ping Chang, (2011): The biosorption of heavy metals from aqueous solution by Spirogyra and Cladophora filamentous macroalgae, *Bioresource Technology*, 102 5297–5304.

Zakaria A. Mohamed, (2001): Removal of cadmium and manganese by a non-toxic strain of the freshwater cyanobacterium *Gloeothece magna Water Research*, 35(18), 4405-4409.

Kolibri 360 GmbH
Ferdinand-Jühlke-Straße 7
99095 Erfurt, Deutschland
produktsicherheit@kolibri360.de

Druck:
CPI Druckdienstleistungen GmbH
im Auftrag der
Kolibri 360 GmbH
Ferdinand-Jühlke-Str. 7
99095 Erfurt